IMPACT

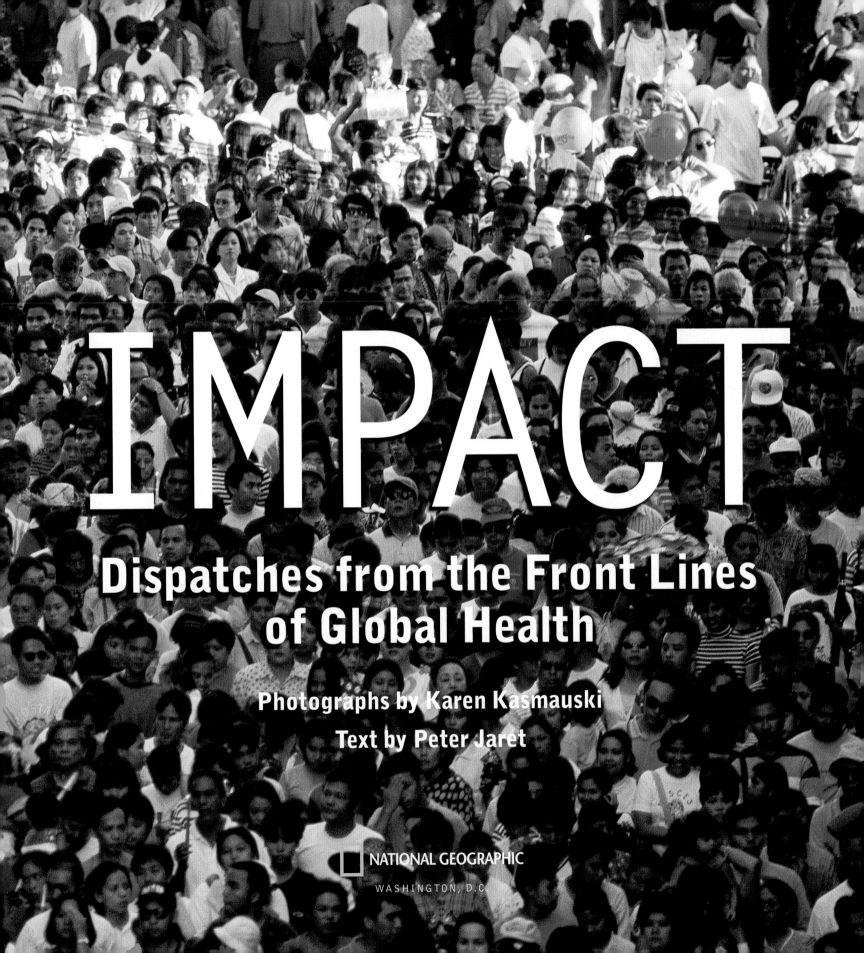

IMPACT

Dispatches from the Front Lines
of Global Health

Photographs by Karen Kasmauski

Text by Peter Jaret

NATIONAL GEOGRAPHIC

WASHINGTON, D.C.

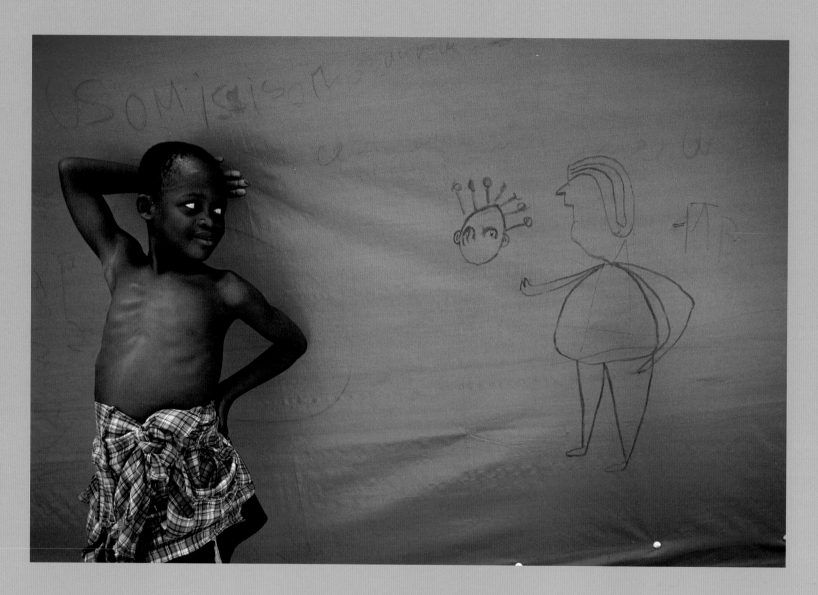

An African boy fleeing civil war may not escape the microscopic legions of disease encouraged by political upheaval. All but invisible, they travel fast and far, carrying lethal consequences. PRECEDING PAGES: Thousands of Filipinos celebrate the annual Feast of St. Niño in Cebu. As global populations rise and gather, people become more vulnerable to transmitted infections.

CONTENTS

Foreword *by Jimmy Carter* 6

Introduction *by Dr. Donald R. Hopkins* 10

I. THE STATE OF THE WORLD'S HEALTH 16

II. SYMPTOMS OF CHANGE 44

Population | Lifestyle | Alzheimer's: A Family Tragedy |
Borders | Political and Social Upheaval | Environmental
Disruption | AIDs | AIDs: One Woman's Story

III. MEETING THE CHALLENGE 164

Clean Water, Safe Food | Empowering Women | Woman to
Woman | Vaccines | Genetics | Healthy Aging

Resources 236

Acknowledgments 237

Index 237

Credits 240

A TALE OF TWO WORLDS

By Jimmy Carter

OURS IS AN AGE OF COMPLEXITY, CONTRADICTION, AND CHALLENGE. As we enter the 21st century, we have wealth and technology unmatched in human experience, and the fortunate few who live in the world's developed nations are almost inevitably propelled toward a future enriched by advances in computers, communication, and life sciences.

But for most of the world's people, the glittering opportunities of the new century are beyond reach. There are more than six billion of us on Earth, and by 2100 we may number ten billion. Most of us will live in urban centers, and many are likely to live short and impoverished lives. Lacking both the wealth and awareness to address problems of life in crowded cities, they will suffer from disease and inadequate food and water.

We face tremendous challenges as populations soar, mostly in the poorer nations, and consumption increases in the industrialized world. We must find ways to lessen the burden on Earth's resources, and we must encourage better stewardship of the planet so that all of us live in a clean and productive environment. The decisions we make in the decades to come will affect not only all of human civilization but also the fate of thousands of species, representing millions of years of evolution.

All too often our fondest expectations are frustrated. Louis Pasteur, father of the science of microbiology and a key figure in the development of vaccines in the 1880s, suggested that humans had the power "to make parasitic maladies disappear from the face of the globe." Yet, since then hundreds of millions of people have died of infectious diseases—tuberculosis, malaria, AIDS, dengue fever, smallpox, cholera, plague, influenza, and scores of others. And after 30 years of discoveries in molecular biology—including DNA cloning, the sequencing of the human genome, and stunning new developments in techniques for human stem cell research—we still face the daily tragedy of preventable human illnesses, some ancient and others new, unpredicted, and even more virulent.

How can we heal our planet and achieve an Earth that nurtures humanity and nature in all their diversity? As individuals, we can act to reduce our risk of exposure to disease and extend care to others. As communities and as nations, we can educate our citizens, legislate ethically and wisely, and support organizations that conduct research and help those who are ill.

Perhaps the most important challenge for the new century is to share wealth, opportunities, and responsibilities between the rich and the poor—for a world where the chasm between rich and poor grows wider will be neither stable nor secure. So far, we have not made enough of a commitment to this goal. Nearly a billion people are illiterate. More than half the world's people have little or no health care and less than two dollars a day for food, clothing, and shelter; some 1.3 billion live on less than *one* dollar a day. At the same time, the average household income of an American family is more than $55,000 a year, with much of the industrialized world enjoying the same, and in some cases an even higher, standard of material blessings.

The best measurement of a nation's wealth is its gross national product (GNP)—the total output of goods and services. The nations of the European Union have set a public goal of sharing four-tenths of one percent of their GNP with the developing world. But the United States and most other rich nations fall far short of this goal. Our contribution must increase greatly if we are to face future challenges to humanity with any real hope of success.

A growing number of private organizations, some quite small, are working to alleviate humanity's problems. During the past 20 years, for instance, The Carter Center has focused more than half its effort and resources on health care in Africa and Latin America. Experts have helped us identify some of the diseases we can hope to eradicate. One is caused by a parasite called Guinea worm. When our Guinea worm program started, more than 3.5 million people had this painful, debilitating disease, most of them in remote parts of Africa where only contaminated water is available to drink. We have now reduced this number by 98 percent. Fewer than 60,000 people remain affected—almost all of them in the war zone of southern Sudan—and we are working hard to address their plight. In our fight against river blindness, another tropical disease, each year we have traveled to villages to treat more than seven million people—preventing the possible blindness they would have suffered simply from the bite of a little black fly.

The success of these efforts reaffirms my faith that this is a time not for despair but for a global commitment to make the most of our scientific knowledge to address the problems of our age. This book highlights the challenges humanity faces in the 21st century: the global fight to control disease, the need to make our food safe and our water clean and to learn to live together fruitfully in megacities. The problems may seem insurmountable, but they are not. We have the tools; we have brilliant, dedicated people to find solutions to our health problems. All we need is a sense of sharing and the will to change. The will can grow from understanding. Once we understand, we can care, and, once we care, we can change. ■

HEALTH: A GLOBAL GOAL

Dr. Donald R. Hopkins

Associate Executive Director of The Carter Center

"The public interest requires doing today those things that men of intelligent good will would wish, five or ten years hence, had been done." —EDMUND BURKE (1729-1797)

WE HUMANS ALL OVER THE WORLD FACE DAUNTING THREATS TO OUR HEALTH, and those threats are constantly changing. We could do much more than we are doing to reduce these threats, and it is in everyone's interest that we start doing so right now. If we fail to seize this opportunity to improve the future of humankind, we do so at our children's peril—all 2.1 billion of them.

Children born in the most advanced industrialized countries these days experience infant mortality rates of 10 per thousand live births or less and can expect to live an average of more than 70 years. Children born in many developing countries, on the other hand, often face infant mortality rates of 150 or higher and have a life expectancy of 50 years or less. Pregnant women may face even greater odds, depending on where they live. The relative risk of dying in childbirth is 50 times higher for mothers in Africa than for those in developed countries. A study in Bangladesh found that about half of all deaths of females between the ages of 15 and 34 were related to reproduction.

In developing countries excessive rates of disease and death reflect the interlocking effects of poverty, infections, insufficient nutrition, and inadequate spacing of pregnancies. Poor people, for example, may not be able to afford necessary preventive or primary health care, even when it is available. Diarrheal disease, respiratory infections, malaria, and AIDS are major killers. In addition to stunting physical development, chronic undernutrition in children often permanently damages their

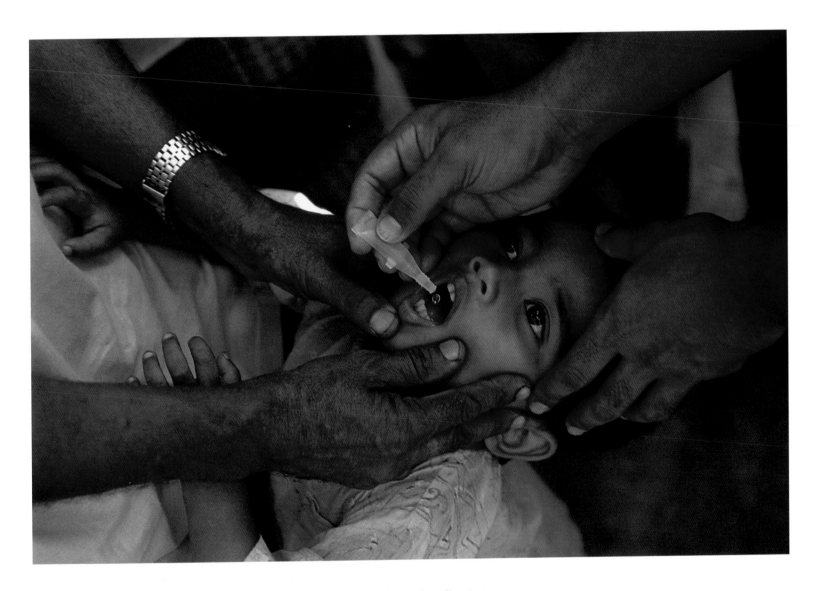

Two drops of oral polio vaccine will protect a
Bangladeshi boy from a scourge that once crippled
millions of children. Global efforts have almost eradicated polio and Guinea worm disease, an African
affliction that can impact whole communities.

ability to learn and think. Frequent pregnancies, with too brief intervals in between, not only compromise the health of mothers and infants but also strain the coping capacities of parents, families, communities, and countries. The resulting increase in population densities helps the spread of such deadly diseases as tuberculosis, cholera, and measles. The cycle of ill health becomes a vicious one: Measles can precipitate acute undernutrition, diarrhea, or fatal respiratory infection; overpopulation and/or undernutrition can increase vulnerability to, and mortality from, measles, tuberculosis, diarrhea, and respiratory infections.

Recently, new influences, including smoking, automobile accidents, and the overuse of refined surgar, have added to these long-standing health risks in the developing world. Exposure to cigarette smoke or to smoke from wood fires in enclosed huts increases the risks of contracting certain types of meningitis and acute respiratory infections (ARI). Studies in Gambia showed that girls under five years old who were exposed to smoke while their mothers cooked on traditional wood stoves suffered a sixfold increase in ARI. Imagine the cumulative impact on mothers in the developing world who have frequent pregnancies; use traditional wood stoves; labor on the family farm, as so many do; and also haul household water—a task that can eat up 20 percent of their time and 9 percent of their caloric intake.

The vicious cycle reaches beyond health itself. A community's general poor health usually means that agricultural productivity and school attendance lag. As an example, a 1987 UNICEF-funded study in Nigeria found that the large numbers of rice farmers suffering from Guinea worm disease (dracunculiasis) resulted in the loss of 20 million dollars a year in unharvested rice. That was only one crop in a small part of one country. River blindness (onchocerciasis) causes populations to abandon large, fertile riverine areas, and African sleeping sickness (trypanosomiasis) kills people and prevents survivors from raising cattle in certain regions of sub-Saharan Africa.

Why should the developed world care? Because these tragedies hurt all of us. Good health, education, and agriculture have been described as the building blocks of society. By impairing one, two, or all three of these building blocks, diseases become serious impediments to development, just as lack of education and poor agricultural techniques are barriers to better health. Better health would help advance many countries toward self-sustaining development and economic well-being, thus making them more self-reliant and stronger partners in the global economy. The recent examples of acquired immunodeficiency syndrome (AIDS), severe acute respiratory syndrome (SARS), and before them smallpox, are powerful proof that the personal health—and with it, the economic health—of everyone on Earth is inextricably intertwined and cannot be untangled.

Some important victories have been and are being won in the ongoing struggle to address these problems. Smallpox is eradicated, even if it hasn't disappeared completely as a cause for concern. Polio and Guinea worm disease will both be eradicated soon, and measles and lymphatic filariasis (elephantiasis), which are more prevalent diseases, may well take their places in the crosshairs of the global public health struggle in the near future. The proportion of children who were properly immunized against several common diseases rose worldwide from less than 5 percent in 1974 to 80 percent in 1990, though it slipped to some 70 percent overall during the 1990s. By 1998, 72 percent of children had been immunized against measles. That disease, which killed more than five million children in 1980, now kills less than a million—still far too many. In Bangladesh, immunizations reduced tetanus infections in newborns by more than 90 percent from 1986 to 1998. The threat of river blindness has been almost eliminated in much of Africa and Latin America and is under increasing attack elsewhere, thanks to programs that first started in 1974.

Meanwhile, the dozen international health experts that constitute the International Task Force for Disease Eradication, based at The Carter Center and supported by the Bill & Melinda Gates Foundation, continue to scan the public health horizon to identify new opportunities for controlling or eradicating diseases. Most diseases can't be eradicated, of course, but many could be controlled much better. Global research and public health programs supported by the massive, targeted resources of the Gates Foundation have, in a few years, started to transform visions of what is possible and expected of public health professionals and organizations. Merck set a stunning precedent with its donation in 1987 of medication to fight onchocerciasis. Since then, GlaxoSmithKline has made similar donations to combat lymphatic filariasis and Pfizer to fight trachoma. DuPont contributed millions of dollars worth of nylon filter cloth and American Cyanamid/BASF the larvicide needed to eradicate dracunculiasis. And Rotary International and Lions Clubs have assumed unprecedented roles in polio eradication and prevention of blindness, respectively.

Other important advances in global health have come from the recognition that health improvements require collaboration with other public sectors to improve such critical areas as agriculture and the availability of safe drinking water. At last, policymakers are coming to understand the importance of a healthy population to a nation's economic well-being.

Advances notwithstanding, a quarter century after the world resolved to implement primary health care as a means of achieving "health for all," little has changed to provide routine, rudimentary health services to most of the world's poor. Our victories have been modest compared with what is needed. We are discovering "new" infections much faster than we are eradicating older ones, and that dynamic is not likely to change. Moreover, many improvements are limited to pilot projects or only parts of some countries. Much more can and ought to be done.

We have the technology to make a much greater impact on global health, and we are acquiring more tools each year: old and new vaccines, tablets that treat many parasites at once, antibiotics, oral rehydration therapy to counter diarrhea, bed nets to protect against mosquitoes, condoms to protect against HIV/AIDS, and more. We are lacking, however, in adequate enlightenment and the money and political will to put our tools to maximal use for everyone's benefit. We need fewer global resolutions and more manifest global resolve to help reduce disease and death wherever we can, as soon as we can, and for as long as necessary.

Current initiatives are piling up. In addition to those mentioned above, we now have AIDS control efforts, STOP TB, Roll Back Malaria, Intestinal Helminth Control, and more. But the nitty-gritty, foundation-building work of improving primary health care services is neglected, even by its most vocal advocates. Some expensive, hard-won gains in disease control are in danger of being rolled back because local primary health care services are too weak to sustain them. Two sad examples of this are African sleeping sickness and yaws. The ongoing training, support, supply, and supervision of peripheral health workers needed to provide routine, prioritized services to fight these diseases is woefully lacking. That frontline health workers are too few, and even those few are commonly ignored, is a failure primarily affecting their communities and countries, but it also means less protection for the rest of the world as well.

Countries need sustained help to come up with programs that address their priority diseases simultaneously, and they need to be held publicly accountable for meeting announced, disease-fighting benchmarks along the way. We have missed some opportunities already. A generation ago, immunizations to control measles were conducted simultaneously with vaccinations to eradicate smallpox, but only in West Africa, while such a combined strategy for polio eradication and measles control was used only in the Americas. Both were successful. The Carter Center is now helping two Nigerian states to combine health education with mass drug administration against onchocerciasis as part of the African Program for Onchocerciasis Control (APOC), with similar interventions against schistosomiasis and lymphatic filariasis. But Nigeria has 36 states and APOC covers 19 countries, each of which has other diseases that also require better control.

Badly needed improvements in public health cannot be achieved in the typical three- or five-year time frame. It would be more realistic for developing countries to seek sustained assistance from developed countries until they can stand on their own feet in the fight against disease. Since the late 1970s, the Centers for Disease Control and Prevention (CDC) has helped 20 Asian, African, European, and American countries develop national programs modeled after its own Epidemic Intelligence Service (EIS). In each country a single, experienced epidemiologist from CDC works for about five years, training local physicians to do routine surveillance and analyses, to investigate suspected epidemics, and to conduct operational studies under local conditions. Simultaneous assistance in upgrading diagnostic laboratory service is sometimes included. Even while they're being trained, however, the trainees are producing epidemiological information that is useful to the ministry of health. Within five years or so, the country has a service that is self-sustaining because graduates of the program help train and mentor new recruits while working in health posts, universities, and public health institutions nationwide. Every country needs some version of such a service, and the whole world would benefit from this. Year after year, we struggle with the consequences of not having such services. Disease surveillance, control, and eradication in Chile, China, and Chad is the world's business, not just a national concern. Microbes recognize our common humanity even if we don't.

We need greatly increased, sustained First World assistance, combined with Third World political will, and a mutual insistence on measuring success or failure by reductions in disease in villages and towns around the world. We need a grand alliance against disease, a sustained war on microbes, including continued research, and real progress in strengthening primary health care for rural and urban populations everywhere.

Mozart almost died of smallpox as a child. How much poorer would the world be if Nelson Mandela had died of measles as a boy? The world is losing potential scientists, statesmen, and artists every day, and we are all the poorer for it.

The microbes have already declared war on us. We need to come together to declare war on them and on other barriers to better health for us all. It's not charity. It's common sense.

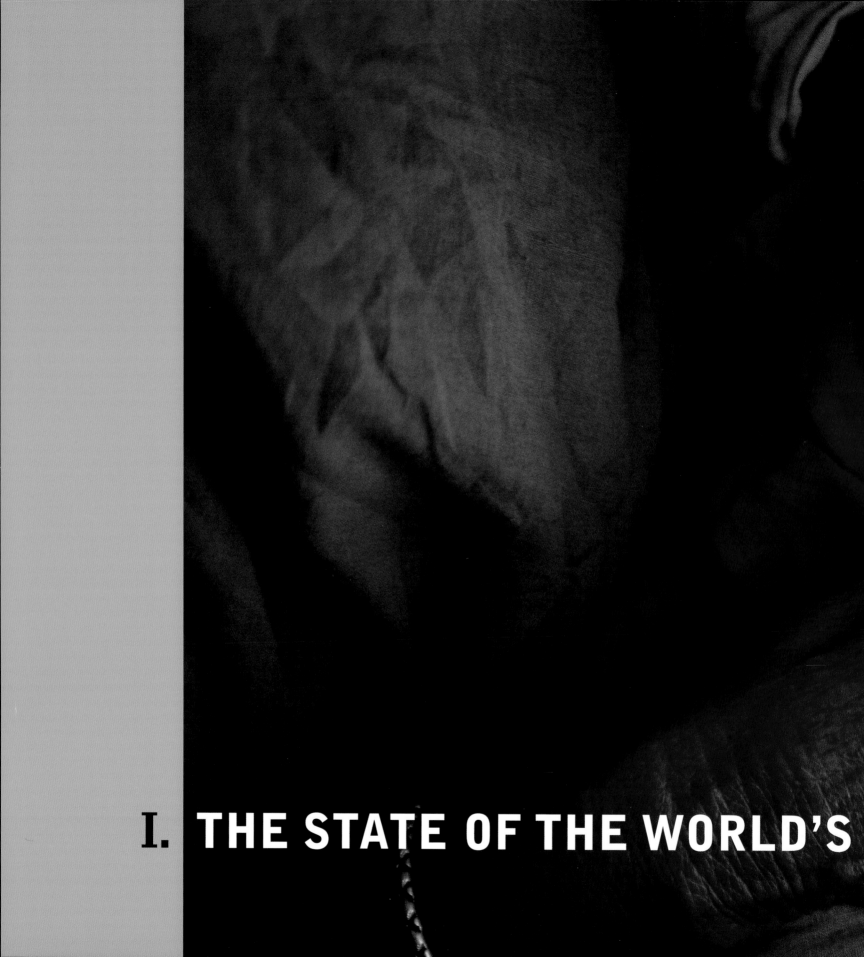

I. THE STATE OF THE WORLD'S

HEALTH

CHAPTER I

THE STATE OF THE WORLD'S HEALTH

In Tanzania, a weary mother nurses one of her children in a refugee camp, exhausted after fleeing the brutal fighting between Tutsi and Hutu. At Lincoln High School in Nebraska, a group of teenage mothers hold their squirming babies in their laps—part of a program designed to help them graduate despite the demands of young parenthood.

In Niger, a small family of nomadic cattle herders sleeps under a bed net, their only protection from mosquitoes that carry malaria. In Macon, Georgia, a mother infected with HIV cuddles with her son, reading him a bedtime story, while halfway around the world, a family in Uganda buries one of its members, the latest casualty of the AIDS epidemic.

The photographs in this book, taken over the past 15 years in almost every corner of the Earth—from the rain forests of Brazil to the crowded slums of Calcutta, among the world's poorest and its most privileged—are an attempt to describe the human condition. Their sweep is global; yet they also allow us to look directly into the faces of individual people going about their lives. Their subject is of one of the grandest and most daunting of all human enterprises: the effort to improve health around the world.

Perhaps no other endeavor captures more fully the complexities and contradictions of humanity. The quest to give everyone on the planet a fair chance at a healthy life involves science, economics, politics, environmentalism, psychology, culture, education, activism, and much more. It's carried out by microbiologists charting changes in a new virus, volunteers bringing polio vaccines into remote rain forests, researchers developing new drugs, statisticians looking for links between diet and health, activists rallying for more funding for health care, and thousands of others. Public health is conducted at the highest levels of government and in the most remote villages, by some of the world's wealthiest philanthropists and some of the poorest people on the planet.

What weaves all these efforts together? The heart of public health is the simple belief that every single human life matters. The muscle of its determination comes from the stubborn conviction that everyone deserves access to the essentials of good

Ten years after the *Exxon Valdez* spill, a footprint
still glistens with oil on an Alaska beach.
PRECEDING PAGES: On the other side of the globe, a
child is born to hope in Bangladesh. The health of
the planet and its people remain interdependent
and the outcomes far from certain.

health, whoever they are or wherever they were born. Its power comes from our ever expanding knowledge of the causes of human afflictions and how they can be stopped.

"WE LIVE IN MUCK AND FILTHE," PROCLAIMED A LETTER TO THE *LONDON Times* on July 3, 1849, written by the city's poor. "We ain't got no priviz, no dust bins, no drains, no water-splies, and no drain or suer in the hole place....We all of us suffur, and numbers are ill, and if the Colera comes Lord help us." Five years later, cholera came. A man waking up in good health, it was said, would be a corpse by sundown. Within a mere 250 yards of the intersection of Cambridge and Broad Streets, more than 500 people died in little more than a week. At the time, the cause and deadliness of cholera was a mystery. But a young physician named John Snow was convinced he knew how to stop it. Drawing crosses on a map where the victims had died, he noticed that his marks were clustered around several of the city's public water pumps, especially the Broad Street pump. Contaminated water, he deduced, was spreading the disease. Snow persuaded the city fathers to remove the pump handles to stop access to the water. And the cholera epidemic began to ebb.

The story of John Snow and the Broad Street pump has come to symbolize the three missions of public health: to detect outbreaks of disease, locate their source, and intervene to stop them. Doctors heal patient by patient; public health is practiced community by community, country by country. Sometimes the tools are medicines or vaccines, sophisticated blood tests or state-of-the-art genetic sequences. But they are just as likely to be commonplace items, the descendants of John Snow's pump handle: storage containers to keep water from becoming contaminated, bed nets soaked with mosquito repellent to keep out malaria-carrying bugs, brochures encouraging people to quit smoking, condoms to prevent the transmission of HIV.

It's been done again and again. Cholera has been banished from most of the developed world. So has polio. In the 1970s, global public health celebrated its greatest victory, and one of the crowning achievements of the 20th century. Armed with a vaccine against smallpox, an army of experts and volunteers fanned out across the world to immunize systematically everyone at risk. Smallpox, a disease that had been dreaded for centuries, was erased from the inventory of human plagues. With each year that passes, millions of new names are added to the roster of people spared sickness and death thanks to the campaign to eradicate smallpox. There is no better example of the sweeping impact that public health can have on the human community.

Yet even as the victory of smallpox was being celebrated, a new and even deadlier disease was silently emerging. It came at a time when many experts thought the battle against infectious illnesses was on its way to being won. The swift and devastating spread of HIV/AIDS was a humbling reminder of just how vulnerable we remain to infectious organisms. More is known about this ingenious virus than perhaps any other pathogen. We understand its modes of transmission and the way it sabotages the immune system in exquisite detail. But that hasn't prevented tens of millions of people around the world from being infected with HIV. As appalling as the toll already is, it is expected to get much worse in the coming decade.

By making people more vulnerable to infections, HIV/AIDS has opened the door to other diseases once thought to be under control in many countries. Most worrisome is tuberculosis. A person infected with HIV runs a 20-fold greater risk of acquiring TB than the general population. As AIDS spreads, so does TB. Today, an astonishing one-third of the world is believed to carry the TB bacterium. Two million people die of the disease annually. Treating tuberculosis has become increasingly difficult as almost invincible drug-resistant strains have emerged and spread. Worldwide, growing numbers of people have strains of TB that can resist almost all current drugs. As with HIV/AIDS, future projections paint a dark picture. Over the next 20 years, a billion people are likely to become infected with TB; 35 million will die from it.

Another disease gaining strength as the result of new and deadly drug-resistant strains is malaria. This ancient scourge continues to kill one million people a year, with 90 percent of its victims children under the age of five. Even when malaria doesn't kill, it takes a tremendous toll. Repeated bouts of malaria hinder children's long-term physical and mental development. In countries where the disease is endemic, it is a contributing factor to poverty, eroding economic growth and discouraging foreign investment. Increasing impoverishment, in turn, makes it harder for hard-hit countries to eliminate the disease, creating a vicious circle that causes even more illness.

Then there's measles. This highly contagious disease can be prevented with an inoculation that costs about twenty-five cents to deliver. *Twenty-five cents.* In use for more than 30 years, it has eliminated the disease in all developed nations. Not so poorer countries, where measles infects 30 million annually, most children, and kills 900,000.

SUCH IS THE PARADOX OF PUBLIC HEALTH AT THE BEGINNING OF THE 21ST century. We are better equipped today than ever before to fight humanity's war against disease. Medical science has identified the causative agents of virtually all infectious diseases and developed vaccines or treatments for many of them. Researchers have deciphered many of the causes of a host of chronic illnesses, from cancer and heart disease to diabetes and arthritis. We've turned medical wonders that would have been unimaginable just 50 years ago—transplanted hearts, artificial knees and hips, reconstructive surgery—into routine procedures.

Yet diseases continue to plague us. New pathogens are constantly arising. Familiar organisms take on deadly new traits. And despite medicine's powerful new tools, germs have gained important advantages. Researchers now worry that our modern world of congested cities and international travel has in fact created a perfect environment for the emergence and quick spread of new infections.

Even when we know exactly how to prevent disease and when we have the means, vexing challenges remain. We possess a powerful vaccine to protect against measles. We've eliminated the disease in dozens of countries around the world. But poverty, political upheavals, and a simple lack of global determination have allowed the virus to flourish in many other corners of the world. We know that doing nothing more than providing a source of clean, uncontaminated water would vastly improve the health of almost half the people of the world, who currently risk serious illness every time they pour a cup of water. But we haven't managed to do that. We haven't even come close.

Recently, even the world's greatest medical triumph, the eradication of smallpox, has been overshadowed by fears that stores of the virus might be used as a biological weapon. The fact that most of the world's population has never encountered this virus only makes us more vulnerable. Bioterrorism is not merely a frightening prospect. Anthrax spores sent through the mail in the U.S. in 2001 spread fear and death—and offered a sobering lesson in how fast and far deadly agents can spread in a complex, intricately interconnected society. The future could hold far worse scenarios. The same advanced microbiological tools that have made it possible to dissect HIV and the SARS virus now make it possible to "weaponize" germs, making them even more efficient at spreading and causing disease than their natural counterparts. It's even possible to create novel new germs. Vaccinations against known pathogens offer some protection, but they also divert much needed resources from the fight against naturally occurring infectious diseases. Improved disease surveillance in the aftermath of the anthrax scare means we're more alert to the first sign of trouble, but we can't prevent bioterrorism.

Yet despite these dangers, the field of public health today is marked by a surging sense of optimism, born of the creation of bold new coalitions of individuals and institutions determined to bring the benefits of an ongoing revolution in medical science to the people who most need them. The message is simple and potent: We have the tools and the know-how to make a real and lasting difference in the health of people around the world. All that's needed is the commitment to make it happen. That commitment comes from recognizing that we are all connected, one to another. It comes from the knowledge that a disease erupting in a remote corner of a poor country can travel around the globe to the world's richest countries. But it also comes from understanding that even the smallest efforts to help can make a tremendous difference in the lives of the people most in need. ■

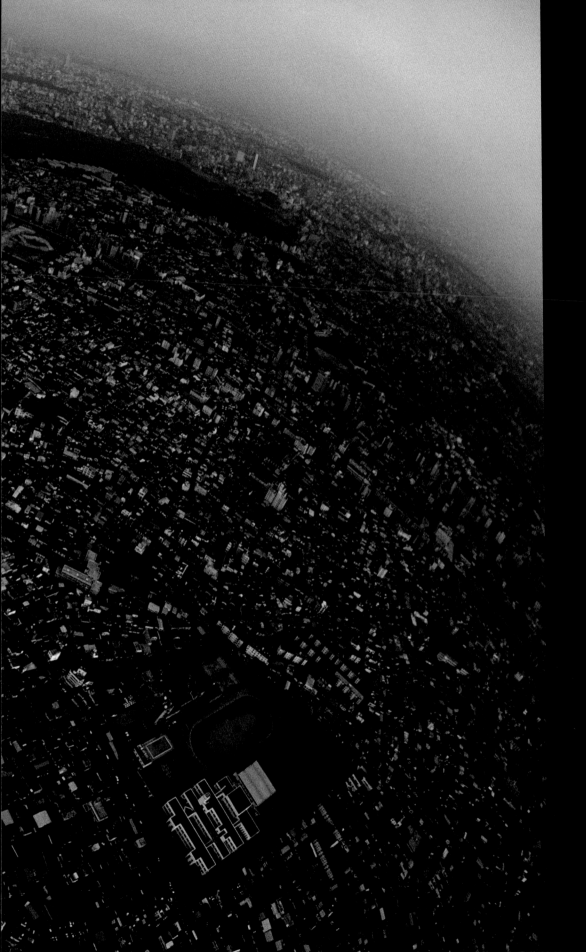

POPULATION

Tokyo manages to pack some 12 million people together in a small space that is mostly clean, healthy, and orderly. Population growth and density can be— but isn't always— a setting for the rapid spread of infectious diseases and social disorder.

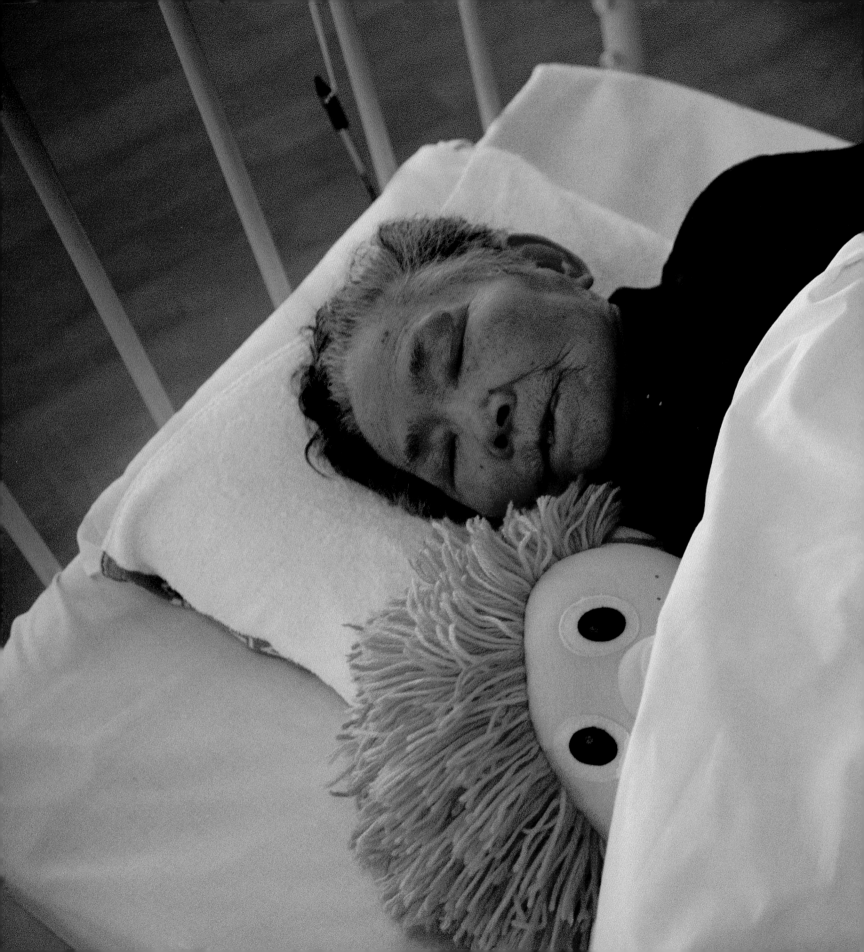

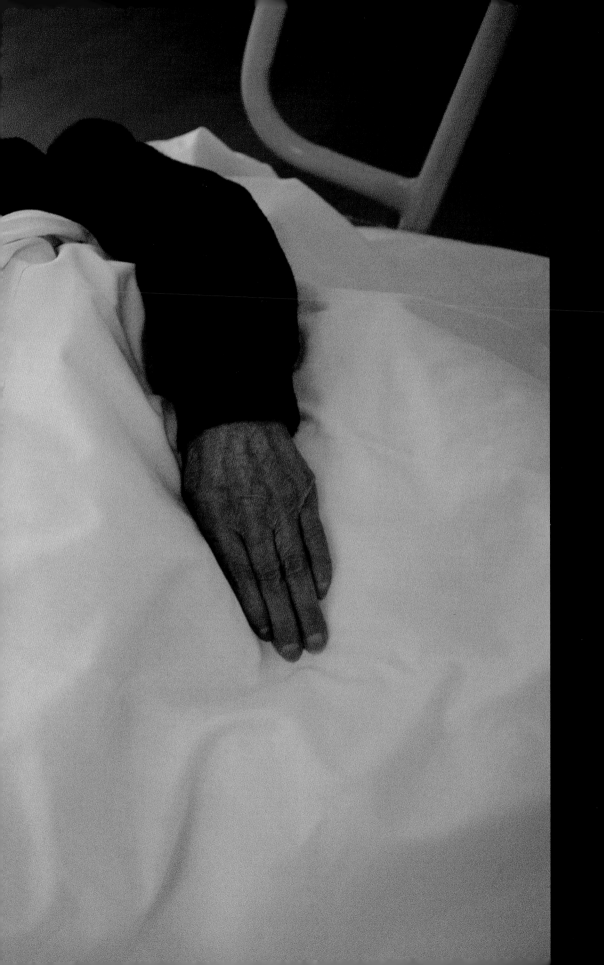

AGE

As the world's population ages, care of the elderly becomes an increasing economic and emotional burden. Will resources allow the old to be treated as humanely as this 87-year-old woman, napping in a Japanese day care center?

OBESITY

Strolling buddies on a San Diego beach salute an unwelcome epidemic—obesity, now endemic in the fast food nation. Nothing to laugh about, obesity increases the likelihood of heart disease, diabetes, and other chronic, life-shortening ailments.

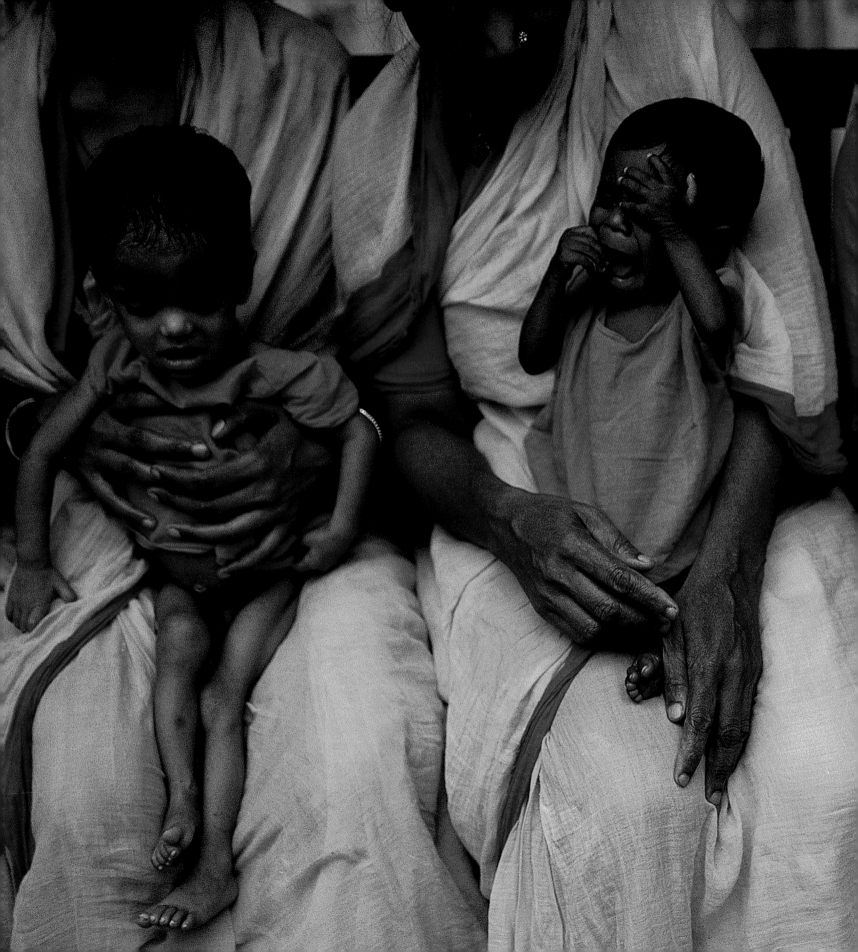

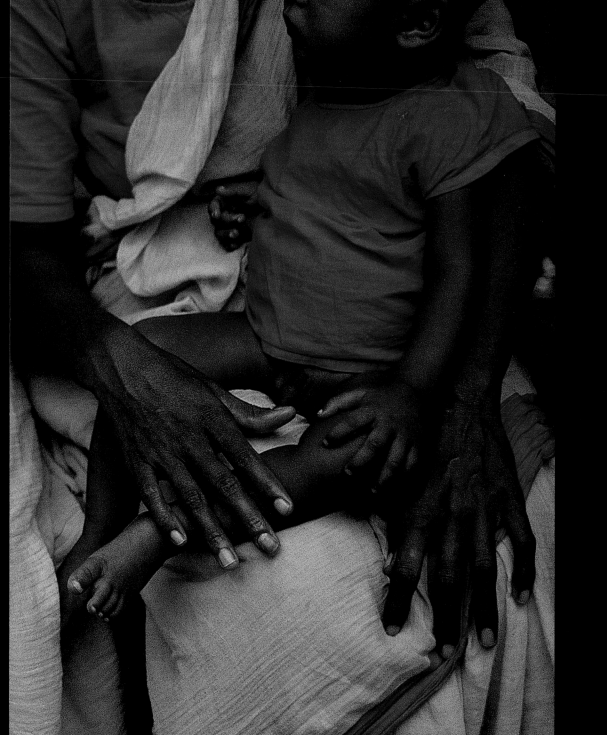

HUNGER

The two Bangladeshi children on the left are malnourished and suffer from diarrhea. The child on the right had the same problems but quickly recovered after eating proper foods for three weeks in a research hospital, where his mother was taught the value of nutrition. Worldwide, malnourished children often succumb to diseases.

Unleashed radiation can be the most potent of killers, but controlled in medical practice it can help save or prolong lives. Doctors at the University of California's Lawrence Berkeley Laboratory treat a woman suffering from a tumor pressing against her spinal cord by bombarding the cancer with heavy-ion radiation.

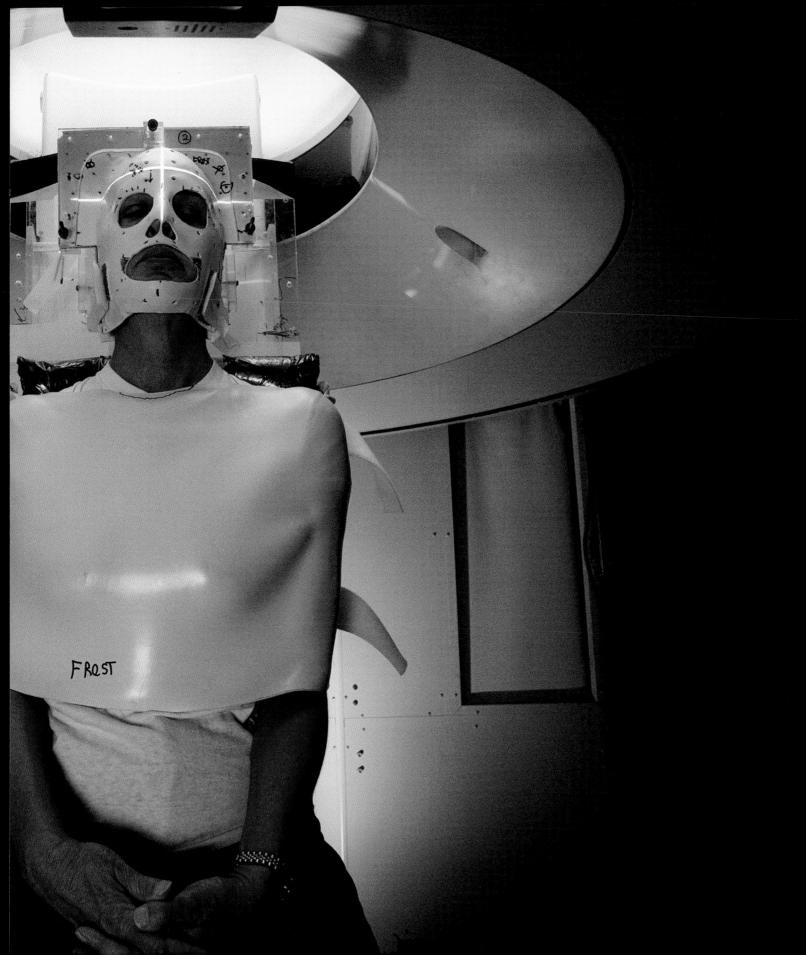

POLLUTION

An outhouse perched over a pond serves as the only plumbing in a Bangladeshi slum. For a community to enjoy health, sanitary sewage disposal and clean water are essentials.

HEALTH CARE

Holding on, a child
visits a Bangladeshi
clinic with her preg-
nant mother, who will
be given a checkup
and told to space
her children. When
taken, such advice
helps mothers and
children live healthier
lives and slows the
planet's runaway
population growth.

AIDS

Mourning her eighth
child to die from
HIV/AIDS, an 80-
year-old Ugandan
woman is left to care
for the surviving three
grandchildren. A killer
everywhere, the AIDS
pandemic has devas-
tated Africa, where
29,400,000 people
have been infected.

VIRUSES

Not even such fear-somely infectious diseases as Ebola and smallpox could survive the chemical scrub-down practiced by researchers at the U.S. Army Medical Research Institute. In the outside world, however, deadly pathogens are not so easily shielded against or disposed of.

EMPOWERMENT

The picture of youth
and hope, a teenage
Masai bride has had
the luck to be born into
a world where females
are being given choices
they never had before.
When women are
educated and allowed
to earn their own
livings, statistics show
that the entire society
benefits with better
health, wealth, and
well-being.

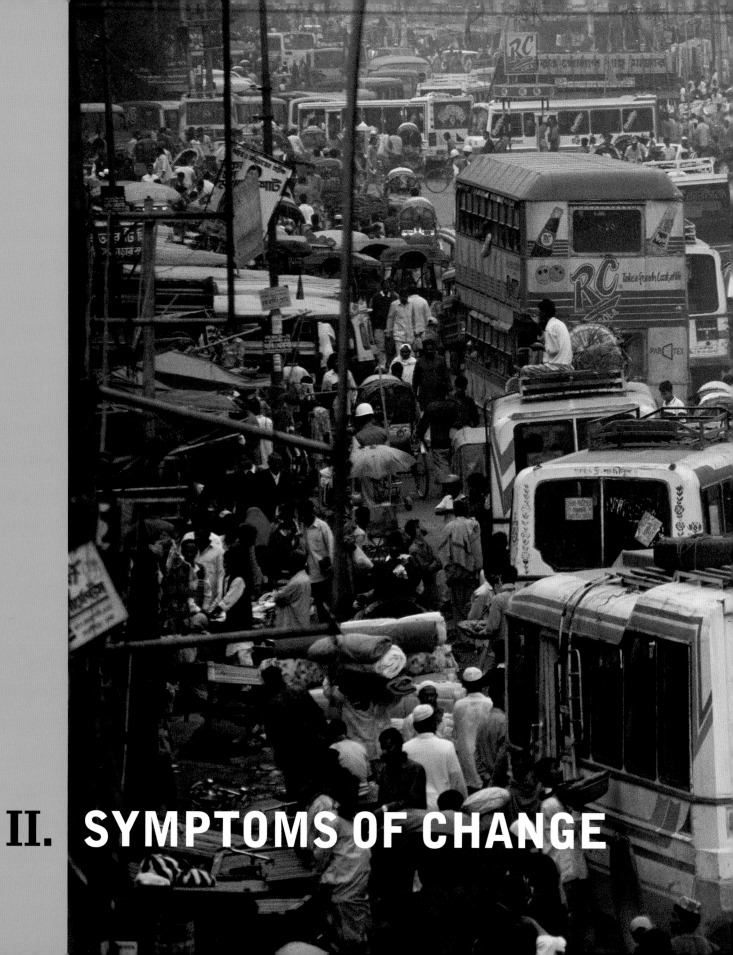

II. SYMPTOMS OF CHANGE

CHAPTER II

SYMPTOMS OF CHANGE

THE CASE MIGHT HAVE BEEN JUST AN ANOMALY. AN AMERICAN MAN, transferred to a hospital in Hong Kong from Vietnam, dies of an unusually virulent pneumonia in late February of 2003. Unexplained deaths like this occur occasionally without raising an alarm.

But within a week, hospital workers at the medical center in Hong Kong where the patient died begin to fall sick with the same symptoms of fever, weakness, and dry cough. Then comes word that a similar outbreak is coursing through the hospital in Hanoi where the American patient had first been admitted.

Experts initially think the cause might be influenza. Flu viruses occasionally leap from birds to people, with disastrous consequences. A month earlier, two people in Hong Kong were diagnosed with a rare form of influenza found in birds. One was dead. In 1997, the same lethal strain infected 18 people and killed 6 in Hong Kong. Influenza experts have long feared that just such a strain would break out and begin spreading.

But a quick analysis shows that the cause isn't influenza. Whatever it is, it's spreading, and spreading fast. Within two weeks, the disease—dubbed severe acute respiratory syndrome, or SARS, by the World Health Organization—is confirmed in 15 countries, including Singapore, Canada, the United States, Germany, and Italy. In Toronto, two hospitals are closed to new patients because of the infection. In Hong Kong, an entire apartment complex is quarantined. Then comes a shocking announcement: The physician from the WHO who first identified the new disease is dead, a victim of SARS. Health officials confirm that an outbreak of a deadly respiratory disease in the Chinese province of Guangdong the previous November was caused by the same mysterious germ.

As researchers scramble to understand the new illness and how it is being transmitted, one thing is frighteningly clear: The world's newest disease has arrived. How far it will spread and how dangerous it will prove to be is anyone's guess.

The emergence of any new disease comes as a shock. It's still hard to accept that new microbial threats to humans can spring into existence seemingly out of nowhere. Yet new or altered infectious organisms are constantly appearing. At least 30 previously unknown human diseases have been identified in the past quarter

Aging populations and low birthrates in developed countries like Japan could mean that the care of elders will eventually overwhelm a younger generation. PRECEDING PAGES: Burgeoning populations in Dhaka, Bangladesh—and many of the world's other large cities—create mounting threats to public health and social stability.

century—diseases for which there is as yet no cure. Some 20 serious illnesses have emerged in new, deadlier, or drug-resistant forms, such as antibiotic-resistant staphylococcus and tuberculosis.

None, so far, has rivaled HIV/AIDS in destructive power. HIV/AIDS appeared at a time when many experts were confident that modern medicine was well on its way to defeating infectious diseases. One by one, disease agents were being identified and dissected, down to their genetic codes. Vaccines and antibiotics were steadily knocking out these age-old enemies. The threat of infectious disease might soon be vanquished, some researchers began to predict.

HIV taught us otherwise. A new plague had been born, as terrifying and destructive as the Black Death of the Middle Ages. As the pandemic exploded, researchers for the first time began to look seriously at the conditions that drive the emergence of new diseases. Their conclusions were deeply troubling. Despite the astonishing technological advances medical science has made in the past half century, we may actually be more vulnerable to the emergence and spread of new diseases than ever before. In many ways, the modern world aids and abets disease agents, making it easier than ever before for them to evolve and spread with deadly speed.

THE FORCES THAT SHAPE THE EMERGENCE AND SPREAD OF HUMAN ILLNESSES are complex and interacting. One of the most fundamental is the force that gives rise to life itself: evolution. Like all forms of life, the agents of infectious disease, locked in a struggle for survival, are constantly evolving. Now and then they undergo mutations that may give them crucial advantages, allowing them to spread faster, infect new hosts, or resist drugs used against them. A small change in a virus's genetic code can turn it from a relatively benign germ into a killer. SARS was born, experts now suspect, when a new variant of a well-known type of virus, called a coronavirus, came to life.

Some germs are continually evolving into new forms. New variants of influenza appear annually. Usually only slight changes occur in the virus's genetic code. But every 15 or 20 years, radically different variants arise, the result of a recombination of two existing influenza viruses. Because these new recombinant viruses are very different from strains that have been circulating, they can elude most people's immune defenses, causing sweeping and often deadly global epidemics, or pandemics.

In the struggle for survival, germs have a few critical advantages. One is their rapid reproduction rate. Bacteria and viruses can double in number in a matter of a few hours. Over the course of a month, a thousand or more generations may be born, creating a speeded-up evolutionary process that allows for rapid changes. Another advantage is sheer numbers. A single virus can invade a cell and begin multiplying, and within days tens of thousands of viral clones burst forth, fanning out to invade nearby cells.

The odds of tuberculosis germs evolving the ability to resist the two most important antimicrobial drugs is exceedingly small. But an astonishing number of TB germs swarm in the lungs of a single patient, all of them multiplying rapidly. When tuberculosis drugs are misused, the chances that a strain will be selected that can resist the assault of the drugs vastly increases. When people infected with tuberculosis are crowded together, as they often are in densely populated cities or in institutions like prisons, the likelihood of transmission increases.

TB isn't the only microbe taking on new power. There's growing worry about strains of staphylococcus that can survive virtually all known antibiotics. Drug-resistant malaria is a serious problem in dozens of countries around the world. And the malarial parasite isn't the only bug evolving ways to elude treatment. The mosquitoes that harbor and transmit malaria have also become resistant to insecticides that previously controlled them.

We're partly to blame. The evolution of new disease threats may be inevitable, but we've also been unwittingly hastening the process. Evolutionary change, after all, doesn't occur in a vacuum. The beaks of Darwin's famous finches change year by year in response to changes in weather and available food. So, too, the agents of disease are constantly adapting to changing circumstances. Bacteria begin to evolve antibiotic resistance only when they encounter antibiotics. Ideally, antimicrobial drugs kill off all bacteria. But if antibiotics are used haphazardly or inappropriately, as they too often are, the drugs will kill off the most susceptible germs and allow a small number of the hardiest to survive. These survivors take over.

The widespread use of antibiotics in farm animals has created even more opportunities for resistant strains to emerge. Deadly influenza viruses can appear when strains of the bug harbored by birds and pigs intermix, creating a new virus, which then leaps into people. China is a hotbed for new influenza viruses, researchers believe, because fowl, pigs, and people often live together in small areas. SARS, also born in China, is now believed to have jumped from exotic wild animals used for meat, such as the civet cat and raccoon dog, into people.

HOWEVER THEY SPRING INTO EXISTENCE, EMERGING GERMS FIND MORE AND more fertile soil to grow and spread in today's crowded and interconnected world. Human population on the planet has soared from only 1.7 billion people in 1900 to over 6 billion today. It could very well double again by the middle of this century. The more of us there are, the more hosts human disease pathogens have to feed on—and the more opportunities to flourish. The problem is made even worse when growing human ranks are concentrated in densely populated cities, as they increasingly are, the result of a relentless migration of people from rural to urban areas over the past century. By the year 2025, 65 percent of the world's population is expected to live in cities—incubators for diseases like measles, polio, and such waterborne infections as cholera.

Not all crowded cities foster disease, of course. Tokyo is one of the most densely populated places on the planet, and one of the healthiest. But wherever poverty and lack of sanitation and clean water exist alongside crowded conditions, as they do in many of the world's poorest cities, urban areas become ideal mixing vessels for microbes.

Surging human numbers also put unprecedented pressure on the environment, causing new health concerns. Experts worry especially about human incursions into remote tropical areas rich in biological diversity, including long-sequestered germs. And in the U.S. and Europe reforestation has caused a resurgence of the spirochete that causes Lyme disease, carried by deer and deer ticks. As people have built more and more houses in and on the edges of the forests, ticks have spread the infection to people. In South America, the conversion of grassland to agricultural fields set the stage for large outbreaks of Junin virus, carried by rodents.

While human diseases have long traveled roads, sea lanes, and trade routes, the ease of modern travel offers emerging microbes quick passage almost anywhere in the world. Bubonic plague is thought to have arrived in Europe from Central Asia in the 14th century by way of the Silk Road. The Spanish brought smallpox and syphilis to the new world centuries ago by ship. Historians have come to understand that diseases carried from one part of the world by explorers and conquerors played a major role in shaping human history.

The difference today is speed. Someone infected with drug-resistant tuberculosis can travel halfway around the world in less than a day. An exotic virus from one continent can leap across the globe and establish a foothold long before the first signs of trouble appear.

It happened in 1999, when an exotic and dangerous virus, researchers now suspect, stowed away onboard a transatlantic flight from Israel to New York City. No one knows if it was carried by an infected bird smuggled into the country or by a passenger infected with the virus. However it arrived, the new invader was quickly picked up by mosquitoes and spread to birds, which serve as a fertile reservoir, vastly amplifying its numbers.

In late summer that year, crows and other birds began dying. Then, at a hospital in Queens, the first patients began showing up with encephalitis. What virologists discovered shocked everyone: The patients were infected with West Nile virus, a pathogen previously unknown in the Western Hemisphere. Year by year the epidemic of West Nile virus has spread up and down the East Coast and westward, causing more than 4,000 serious infections and some 280 deaths in 2002 alone. It has pushed north into Canada and south into Central America, where it continues to expand its territory.

The world's newest infectious disease, SARS, also took flight via international travel. In just one example of how fast and far a new germ can move, officials have documented a case in which a single traveler, unaware that he was infected with the SARS virus, took seven separate international flights over the course of a week, ranging over

Asia and Europe, before being hospitalized with severe acute respiratory syndrome.

Trade also provides new avenues for the movement of microbes. Rats stowed away onboard freighters carried hantaviruses around the world. *Aedes albopictus*, the Asian tiger mosquito, arrived in the United States, Brazil, and parts of Africa in shipments of used tires from Asia. Since 1982, this dangerous disease transmitter has found a home in at least 18 states in the U.S. In the summer of 2003, the first known outbreak of monkeypox virus occurred in the Western Hemisphere. The exotic virus, which is endemic in the rain forests of West Africa, may have arrived via imported Gambian rats sold as pets in some parts of the U.S. The virus then spread to prairie dogs, researchers speculate. Within two weeks of its discovery, more than 50 human cases were reported.

As dangerous as trade in exotic animals is, the worldwide shipment of food offers an even wider network for germs to spread, since many pathogens flourish in contaminated meat, milk, and other foods. In the U.S., contaminated meat shipped out to dozens of states caused large and deadly outbreaks of *E. coli*. When the United Kingdom experienced a massive epidemic of foot-and-mouth disease in farm animals such as cattle and sheep, a worldwide ban on beef from the U.K.—along with the destruction of millions of animals—managed to prevent the disease from spreading. But the risk of such a catastrophe is ever present, as the recent discovery of mad cow disease in Canada attests.

Almost anything that links us, in fact, can also serve as a conduit for disease. Blood transfusions and organ transplants save thousands of lives every day. Unfortunately, dangerous pathogens can lurk in blood and transplanted tissue. In the early years of the AIDS epidemic, thousands of people were infected via transfusions. Several forms of hepatitis, a sometimes lethal infection of the liver, can be spread through blood transfusions and organ transplants. The U.S. Centers for Disease Control and Prevention recently found that West Nile virus can be transmitted the same way. Screening tests are now in place to prevent such transmission.

Far more dangerous than the use of medical procedures like blood transfusions, however, is the breakdown of medical care and surveillance systems designed to detect the spread of new infections. In parts of Africa, deteriorating conditions have allowed diseases like malaria and tuberculosis to blossom anew. When the deadly Ebola virus erupted in what is now the Congo in 1995, poor health care practices spread the infection from the hospital staff to their families and other patients. Almost 20 weeks passed before news of the outbreak was reported to the international health community—five months during which the ferociously deadly germ could easily have spread. The same scenario is almost certain to be repeated. In many parts of Africa, a National Academy of Sciences report warned in 2000, "infectious disease surveillance is nearly nonexistent, and emerging infections frequently go unreported."

Africa isn't alone. In the aftermath of the breakup of the Soviet Union, inadequate medical care turned Russia's prison system into an incubator of tuberculosis. Three hundred thousand otherwise healthy Russians were being incarcerated and another 300,000 were being released every year, most of them having been exposed to TB during their incarceration. Ten percent developed the disease. Eighty percent carried the germ out into the world, where it could spread to others.

FAR FROM BEING SUBDUED, INFECTIOUS AGENTS REMAIN MAJOR KILLERS. Worldwide, six infectious illnesses—influenza, HIV/AIDS, diarrheal diseases, tuberculosis, malaria, and measles—account for almost half of all deaths between birth and age 44. A rogues' gallery of other less well-known agents, from dengue fever virus to Guinea worm, add to the havoc.

Yet pathogens are only one of the pressing public health challenges we face at the beginning of the 21st century. Environmental degradation, one consequence of surging human populations, directly threatens the health of millions of people, posing what is likely to be a growing problem. Thousands of wells were constructed in Bangladesh to provide a reliable source of clean water; many of them now turn out to be contaminated with dangerous levels of naturally occurring arsenic. As many as 57 million people are drinking water that could be poisonous. In Chernobyl, the site of the world's worst nuclear accident, the danger is radiation.

Doctors there monitor residents for early signs of leukemia or thyroid problems. In many of the world's poorer countries, agricultural workers are exposed to potentially harmful levels of pesticides and herbicides. The destruction of vast tracts of forests has disrupted watersheds, allowing once clean water to become contaminated.

Too often, in fact, we are our own worst enemies. Today we know that aspects of lifestyle, including lack of exercise, smoking, and poor diet, are the principal contributors to cardiovascular disease, the leading cause of death in the world's rich countries. The growing use of tobacco around the world represents a health crisis of its own. More than a billion people use tobacco worldwide. An estimated four million people die annually as a result of tobacco-related illnesses. As tobacco companies intensify efforts to sell their deadly products to people in developing nations, that number is almost certain to rise. By the 2020s, according to WHO predictions, smokers will number ten million—70 percent of them in developing countries.

The world's richer countries face another epidemic of a very different sort: diseases linked to obesity. Ironically, the problem of obesity is caused in part by our own success: an abundant supply of relatively cheap food and the development of labor-saving devices that have reduced the need for physical labor. The epidemic of fatness is most advanced in the U.S., where an estimated 127 million people are overweight, 60 million are obese, and 9 million are severely obese. But the problem is beginning to appear in many other developed countries, from England and France to Australia and countries in the Middle East. Diabetes, one of several serious illnesses linked to obesity, is already showing up at earlier ages in children and teenagers who are overweight. An epidemic of the disease will place an enormous health care burden on even the wealthiest countries.

TAKEN TOGETHER, THE MULTIPLE AND VARIOUS THREATS TO HEALTH AROUND the world can seem overwhelming. Sometimes they are. A massive global effort to make clean water available to people around the world, launched in the 1980s, fell far short of its goal—though it did help launch a global campaign to eradicate Guinea worm disease. During that same decade, an international coalition set its sights on reaching at least 80 percent of the world's children with polio and DPT vaccines. Four out of five children were vaccinated, but the campaign didn't meet its more fundamental goal: to give individual countries, even the poorest, the means to continue immunization programs on their own. With no infrastructure in place, immunization rates rapidly declined. Preventable childhood illnesses once again began to afflict new generations of children.

Tragically, the global campaign to stem the spread of HIV in many parts of the developing world has also come up short. Every day, 14,000 more people are infected. The fastest growing rates of new infections are in some of the world's most densely populated countries, raising fears of another vast wave of new illnesses. Within ten years, 100 million more people could become infected. The failure to stem HIV/AIDS can create a sense of helplessness and disillusionment. As the 2000 report of the Global HIV Prevention Working Group acknowledged, "The global community often behaves as if a massive expansion of HIV/AIDS were inevitable, as if the world has little choice but to watch anxiously and hope that the epidemic eventually burns itself out."

That sense of inevitability and helplessness represents yet one more factor that can serve to spread disease—one potentially as dangerous as any emerging virus or new vector. Human disease flourishes not only where there is poverty and instability but also where there is little or no hope. Where people have little stake in the future, for example, it is difficult to persuade them to take precautions against HIV infection. When a family can do almost nothing to protect its children from even the most easily prevented waterborne infections, it is impossible to begin building the groundwork for good health. And when people in the developed world believe there is little they can do to improve the lot of those in the poorest countries, then indeed little will be done.

Fortunately, disillusionment is being countered today by a variety of innovative public health initiatives—some as modest as a radio soap opera in Botswana designed to deliver subtle messages about HIV/AIDS, others as ambitious and sweeping as the campaigns to eradicate polio, Guinea worm, and lymphatic filariasis. They show a renewed determination to confront the world's most challenging health problems. That determination is the necessary step toward a real and lasting difference. ▪

POPULATION

Over the next half century, the Earth's population is expected to swell from the current 6 billion people to as many as 11 billion. That extraordinary increase in human numbers will impact health in many ways. For infectious organisms, our swelling ranks offer fertile ground. The more of us there are, the more hosts we provide for an array of pathogens. The surge in population is accompanied by vast movements from the countryside to the city, creating human habitations of unprecedented density—and the ideal conditions for infectious diseases like measles and influenza to spread. In many of the world's poorer cities, rapidly growing populations overwhelm already fragile infrastructures for water and sanitation. The result: more opportunities for diseases like cholera or hepatitis A to find a home.

Growing populations will also place an enormous burden on health care systems that are already falling short of meeting existing needs. Developing nations today account for 98 percent of the population growth on the planet. Eight countries are expected to account for half the population increase in the next half century, according to a recent UN report, including India, Pakistan, Nigeria, China, Bangladesh, Ethiopia, and Congo.

Surprisingly, that list also includes the United States, where population is expected to climb from 285 million today to 409 million in 2050. The U.S. is an exception among developed countries. Many of the rest have seen dramatic declines in their birthrates. In some, fertility rates have fallen below 2.1 babies per woman, the rate that keeps the population stable. Europe's population is expected to shrink from 728 million people today to 632 million in 2050. Even among developing countries there are signs of a slowdown in population. One reason is a grim one: the growing toll of HIV/AIDS. The UN recently revised downward its population forecast for 2050, in part because of deaths caused by the epidemic. Half a billion people who would have been living on the planet in mid-century won't be alive or won't have been born because of AIDS.

But there is a heartening reason why the planet's population explosion is being tamed: greater economic opportunity and expanding personal choice in family planning. In Bangladesh, the 1990 birthrate was nearly 5 children per woman. Unchanged, that would have meant a doubling of the population by 2015. But that dire forecast didn't come to pass. The birthrate has fallen to only 3.3 children per woman—an astonishing change in a short period of time. Across the developing world, fertility rates now average 3 children, down from 6.5 a century ago. In the near term, growing human numbers will pose serious challenges. But current trends suggest that the world's population will begin declining in the second half of this century—offering both new opportunities and a set of new challenges.

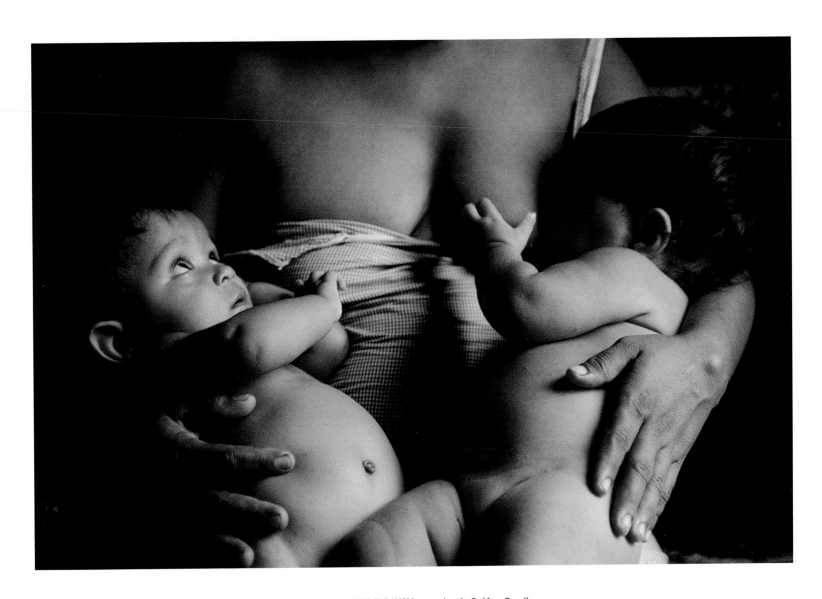

TAKING NATURE'S WAY, a mother in Belém, Brazil, nurses her twins. Mother's milk, full of antibodies, gives babies a healthy start, yet for years women in developing countries were encouraged to use formula powder that needs mixing with water. Result: Children suffered more waterborne illnesses—and they missed out on the natural antibodies in breast milk.

PACKED APARTMENTS in Ho Chi Minh City and tract housing in a San Diego suburb both speak to a growing global population. Of the developed countries, the U.S. is the only one with a population boom rivaling that of developing nations.

SIMPLE BREAKFAST:
Masai children in Kenya sip tea before heading off to school. Their father, a pastoralist, has three wives, who have each had a number of children. With limited cattle—the traditional basis of Masai wealth and standing—to go around, not all the children will inherit enough cows to support them. Planning for that eventuality, their mothers craft jewelry for export to Europe, using their earnings to school their children for a new kind of future.

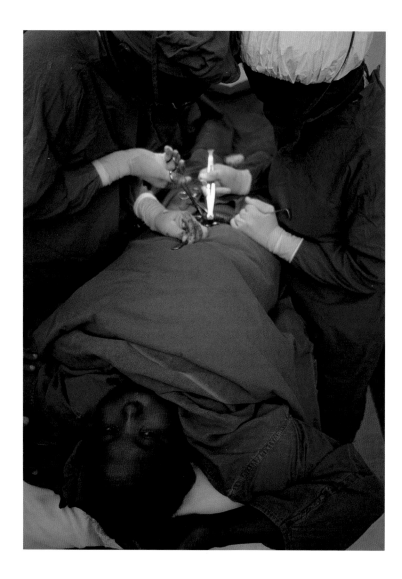

FAMILY PLANNING, a lifesaver: In Kenya a woman undergoes sterilization (left). Fewer pregnancies mean healthier, better-cared-for children. In developing countries, children weaned as new infants arrive are particularly vulnerable to health problems, as they must compete with older siblings for the limited food available.

TEEN MOMS IN Nebraska learn to care for their infants—and their own futures—in a program that aims to keep premature parents in high school. The U.S. has the highest teenage birthrate in the developed world. When children have children, the path toward maturity, economic success, and long-term family stability requires commitment and training.

LIFESTYLE

People in the world's richest nations enjoy lives of leisure and plenty that most of the rest of the world can hardly imagine. But lately we've begun to understand that even the seeming benefits of modern civilization can threaten health.

In the 1950s, when epidemiologists began to investigate a shocking rise in premature heart attack deaths in the U.S. and parts of northern Europe, the clues led to an unexpected villain. Diets laden with saturated fat, researchers learned, can increase blood cholesterol, which in turn raises the danger of clogged arteries and heart attacks. The lack of fiber in many modern diets exacerbates the problem. Inactivity is another culprit. The human body, we've come to understand, is built to be active. Sedentary lifestyles dramatically increase the risk of obesity and its consequences, including cancer, adult-onset diabetes, and heart disease. The percentage of Americans who are obese has doubled in the past 20 years. The same disturbing trend is seen not only in other developed nations but also in more and more developing countries.

The fact is, *how* people live—their diets, physical activity patterns, tobacco use, the quality of their friendships, alcohol consumption, the pressures at work and home—have a profound impact on health, and not only for chronic diseases like heart disease and diabetes. The course of many infectious illnesses, including HIV/AIDS, is shaped by the choices people make.

Unfortunately, many lifestyle-related diseases are on the rise, especially in developing countries. The causes are complex. Labor-saving technologies are doing away with the need for physical activity in everyday life. Traditional dietary patterns, often based on a healthful selection of locally produced foods, are being disrupted as cultural and family patterns change. Global companies are in the business of aggressively marketing their wares, which too often include tobacco products, high-calorie processed foods, and other fixtures of Western culture. Meanwhile, the rise of global media means that the world's poorest people can see how the most advantaged live—and aspire to live the same way. The result: Smoking is on the rise in many developing countries. So is obesity, heart disease, diabetes, and a host of other plagues related to modern lifestyles.

Fighting this newest epidemic won't be easy. So far, efforts in the U.S. to stem the rising tide of obesity have been largely futile. Finding ways to encourage people to adopt more healthful lifestyles requires a keen understanding not of the nature of viruses or bacteria but of something even more challenging, it seems: the complexities of human nature.

A DRAG on healthful living the world over, smoking tobacco is legal, relatively cheap, addictive—and often deadly. For the world's poor—like this Filipino sugarcane worker—it's still an irresistible and affordable luxury in a long, hard day.

SHAPE UP, THEN ship out is a standing order for U.S. Navy recruits in San Diego. Most Americans know that regular exercise, and a healthful diet will promote longer life. Yet obesity has become a national—and increasingly inter-national—epidemic, with serious health consequences.

SOFT DRINKS AND TV have inundated the globe. Neither offers much nourishment, and nighttime TV watchers, like these Thai girls, are at additional risk, since they're without the protection of netting that helps guard against bites from malarial mosquitoes.

A LEG UP on better health encourages women in San Diego to "Jazzercise." Regular exercise helps keep weight down and promotes good respiratory and cardiovascular function. For many, working out with a group boosts motivation and improves results.

ALZHEIMER'S: A FAMILY TRAGEDY

Ina was in her late 40s when the first symptoms appeared—forgetfulness, difficulty remembering even simple, daily tasks. A decade later, when these pictures were taken, the disease was taking its inexorable toll, and the worst was yet to come.

Striking people under age 65, early-onset Alzheimer's accounts for about 15 percent of the disease's total victims. Whether it occurs as early-onset or later in life, the degenerative march of Alzheimer's sweeps families, as well as patients, along in its wake. As demographics change and the elderly become a greater proportion of the population, whole societies could be dealing with the consequences of a rising tide of the disease.

Researchers have implicated eight genes in the onset of the disease, but with no cure yet available, children of Alzheimer's victims might be reluctant to be tested for their genetic propensity for the disease, if a test were available. Ina's own daughters had very different opinions. "If I get Alzheimer's, I get it," one said. "It would be irresponsible not to [be tested]," another felt, adding, "I want to start a family." The third daughter worried, "What if an insurer learned the test was positive and decided not to pay for treatment."

Ina was lucky to have a caring family, willing to help her cope with the disease as long as possible in her own home. But such caring comes at great cost to the family members, who must rearrange their own lives to deal with the increasing helplessness of their loved one. When relatives live far away, care of any kind becomes fraught with problems. Too often, an aging spouse must shoulder all the burden alone. With genetic research now on the verge of a new era, maybe that burden will soon be lifted.

DRIFTING IN THE FOG of early-onset Alzheimer's disease, a Baltimore woman named Ina is gradually losing her ability to cope with daily living.

LOVING CARE but no cure is the outlook for the most fortunate of Alzheimer's patients. Ina's husband helps her organize her day and medications. Medical professionals consult with Ina and her daughters, but the prognosis is not good. Whatever the cause of her dementia—and it is most likely genetic—effective treatment for Alzheimer's has not yet been found.

EVEN SIMPLE TASKS—remembering to let in the family dog, dressing in the morning—become increasingly difficult for Alzheimer's sufferers. As the disease progresses, aging accelerates, leaving the victim as helpless as a young child.

THE FUTURE for
Ina's daughters rightly
concerns them.
Could they inherit
early-onset
Alzheimer's? If a
diagnostic test were
available, should they
take it? Would insur-
ers pay for treatment
or refuse it? Answers
are not yet available.
The more common
form of Alzheimer's
affects people in their
70s and 80s, and
as the population
of developed coun-
tries continues to
age, dementia will
likely become more
widespread.

BORDERS

Never in human history have disease agents been able to travel so far so fast. A passenger unknowingly infected with malaria, drug-resistant tuberculosis, or SARS can go from Nairobi to London or Hong Kong to New York in a matter of hours. From 1984 to 1994, the number of Americans traveling outside the country doubled. Worldwide, a billion people cross international borders every year.

Travelers aren't the only threat. Increasing international trade has opened up more and more routes for infectious diseases. In 1992-93, a large outbreak of *E. coli* was traced to contaminated meat shipped to restaurants across the western U.S. When a devastating outbreak of foot-and-mouth disease swept England, more than four million animals were slaughtered and imports of British meats banned in order to prevent the organism from spreading outside the U.K. More recently, the discovery of mad cow disease, or bovine spongiform encephalopathy, led to the slaughter of 400 animals in Canada and a temporary worldwide ban on the import of Canadian beef.

When the SARS epidemic began to spread, several Asian countries began questioning travelers at international airports about signs of fever and other indications; those with suspicious symptoms were quarantined. Along the U.S.-Mexican border, border patrol agents screen people for TB. Still, two million people are believed to cross into the U.S. each year without being tested for diseases, many potential carriers of tuberculosis. Each year, hundreds of cases of malaria and dengue fever are also diagnosed in the U.S. in travelers coming from affected countries. These viruses could be picked up by resident mosquito populations and begin to spread—exactly what happened when West Nile virus was introduced into the Western Hemisphere in 1999.

Diseases can jump borders in even more sinister ways, as anthrax mailed through the U.S. post in 2001 chillingly demonstrated. Experts fear that biological agents could be used again in terror attacks across borders.

Yet, although borders can't stop infectious diseases, they can serve as outposts for information. In the border regions on the Indian subcontinent, governments are working together to provide basic preventive health care and treatment to reduce the toll of malaria, TB, HIV/AIDS, and other illnesses. The U.S. Centers for Disease Control and Prevention is collaborating with Mexico to improve surveillance and treatment of several key diseases. Such programs use borders to break down the boundaries that often separate us, improving access to basic health care for people wherever they live.

GERMS FLY FREE in busy airports like Tokyo's and can circle the world in a day. In 2003 severe acute respiratory syndrome (SARS) flew from Hong Kong and in three months infected 8,403 people in 29 countries, killing many hundreds.

FLU FACTORIES, domestic duck farms in China, where animals live in close association with people, create favorable conditions for transmission of viruses to humans. That is how new strains of influenza typically arise, some especially virulent. In 1918 a new strain commonly called Spanish flu emerged and swept the globe. In 16 months it killed 20 million people—almost one percent of the world's population—a fatality figure higher than that from all the battles of World War I. Vaccines have reduced the threat of such pandemics, but even now, on average some 20,000 people in the U.S. die annually from influenza.

A PIG'S PART in the spread of influenza may be more than simply acting as a carrier for swine flu. Pigs can carry both human and bird viruses, and those viruses may exchange RNA, giving rise to new strains. In a Chinese slaughterhouse, scientists check for just that.

"SENTINEL CHICKENS," bellwethers for the spread of West Nile virus, are monitored for signs of the disease. Once confined to Africa, it appeared in the Middle East, then in 1999 traveled to New York City. By 2003 the virus had spread across North and Central America. Right: Participating in a university study, a student volunteer gets a dose of the common cold—like SARS, a coronavirus.

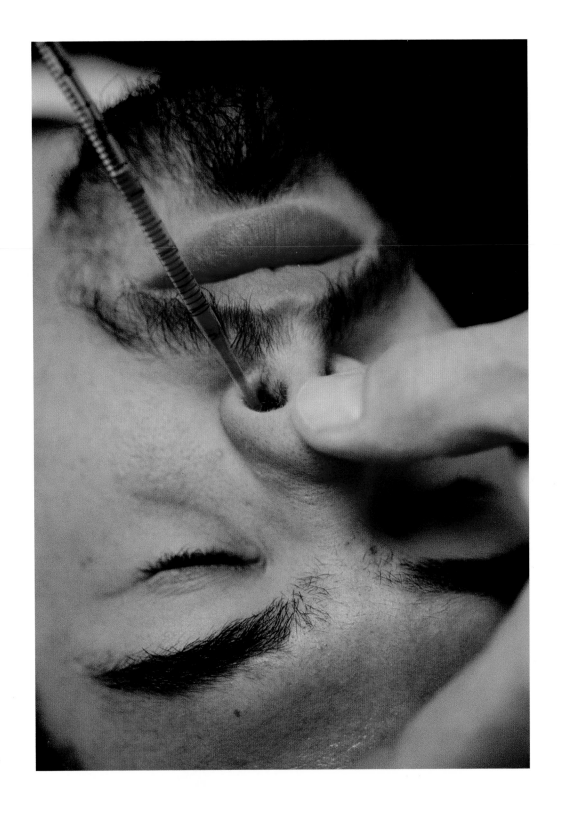

ANTICIPATING THE worst: In a drill, the U.S. Army medical evacuation team practices how to handle a victim suspected of being infected with an extremely contagious and dangerous pathogen—whether naturally or by an act of bioterrorism. Isolated in a sealed, self-contained life-support litter, the victim is transported to a medical facility for observation and treatment.

HANDLED WITH EXTREME CARE, smallpox
vaccine is produced in a laboratory outside
Washington, D.C. (opposite). Though smallpox has
been eradicated, its living victims, like this man
from Niger, bear the scars, blindness, and deafness
millions once suffered.

POROUS BORDERS
Between densely populated Mexicali, Mexico, on the right and wide-open Calexico, California, on the left, people— and diseases—are easily exchanged. Mexican migrants can import diseases less common in the U.S., while visitors to Mexico can carry infections to remote areas whose residents have little immunity.

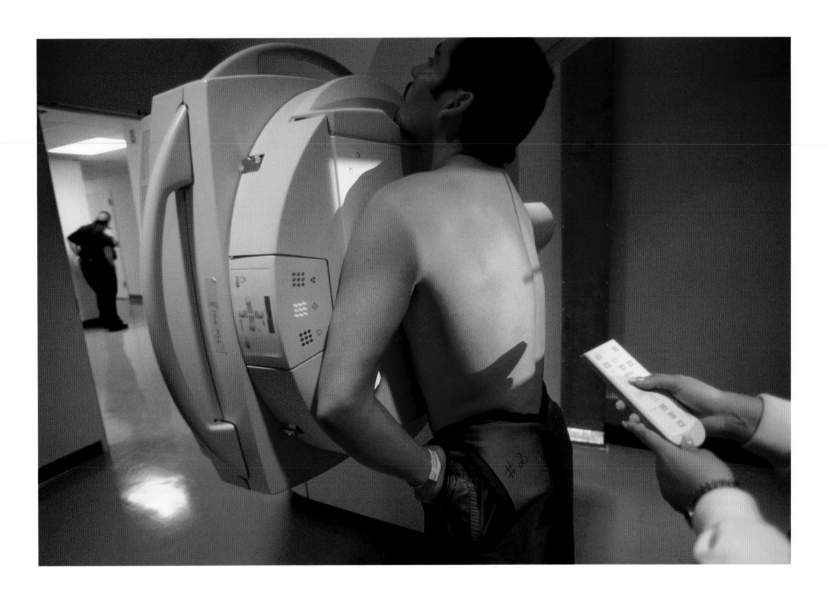

BREACHING BORDERS, detainees like these men can find themselves on the wrong side of the law. With tuberculosis a growing concern along the U.S.–Mexican border, frequent detainees are given chest x-rays to check for the disease.

STREET-SMART aliens from North Africa know
that if they can get to Italy's Lampedusa Island,
they have a good shot at making their way to
mainland Europe. After a cursory health check,
detainees are marched through town, given pocket
money, and put on a ferry to Sicily. Once there,
they're given 48 hours to leave Italy. What many
do instead is disappear into Europe, becoming part
of a growing refugee tide that strains social and
medical services, frays local tempers, and con-
tributes to social unrest.

POLITICAL AND SOCIAL UPHEAVAL

Germs and diseases thrive on chaos. Again and again, the world has witnessed the same tragedy: conflict, political upheaval, or social unrest followed by devastating outbreaks of illnesses. In Zaire, an uprising of troops and a rampage of looting in the capital all but destroyed the economy and much of the infrastructure. Aid groups were forced to withdraw from the country and the World Bank suspended its projects. Diseases like malaria and HIV/AIDS went on a rampage of their own.

In Rwanda in 1994, civil war led to the displacement of some three million people—and the deaths of 100,000 in refugee camps, mostly from cholera and dysentery. Such outbreaks of waterborne diseases are a common occurrence in camps for displaced persons, where crowded conditions and the lack of sanitation and medical care only make it easier for infectious diseases to spread.

War by its nature is catastrophic, of course. But the specific consequences of modern warfare are new and deeply troubling. According to the State of the World's Mothers 2002 report, the percentage of civilians killed and wounded in hostilities has risen from 5 percent in 1900 to 90 percent in the most recent conflicts. Many are women and children. The dislocation that accompanies and follows conflict is often even more devastating than outright hostilities. At least half of all deaths among young children during conflicts are the result of diarrheal diseases, measles, acute respiratory infections, and other illnesses that could have been prevented or treated in the absence of war. HIV has typically spread fastest in countries caught up in the turmoil of civil war or other political dislocations. Many of polio's last holdouts are countries where strife and political instability have made it difficult or impossible for vaccinators to reach all children at risk.

Economic and cultural upheaval can also impact health in complex ways. After the breakup of the Soviet Union, life expectancy fell as alcoholism, suicide, and a rising tide of infectious diseases such as tuberculosis took their toll.

The lesson is simple, but its consequences pose perhaps the most profound challenge of all to efforts to improve global health. Providing even the most basic health care depends on a relatively stable political and economic infrastructure. Without that, programs to vaccinate children, treat drug-resistant cases of TB, or prevent HIV/AIDS are always in jeopardy. International aid and relief agencies can address the urgent needs in countries in crisis. But the ultimate goal remains establishing a functioning and secure health system. And that, in turn, requires a functioning and secure civil society—something that many countries around the world still lack.

SOVIET LEGACY In the days of the U.S.S.R., Uzbekistan's Aral Sea was heavily polluted with pesticides and badly depleted to irrigate cotton fields. The result: dust and salt storms, infant anemia, and tuberculosis.

A FEEBLE STREAM feeding the Aral trickles past dunes where river bottom once was, while desertification claims the cotton fields once irrigated by Aral water. Blowing salt and sand caused by the sea's demise contribute to respiratory problems for schoolchildren, whose lung capacity is regularly checked in a program that is sponsored by Doctors Without Borders.

AN OLD ENEMY
once counted out,
tuberculosis has risen
again as a serious
contagion, particular-
ly in Russia. The prob-
lem started in
Siberian prisons, after
the breakup of the
U.S.S.R. When
inmates being treated
for TB were released
while still infected,
they found they had
no access to medica-
tion once they were
free. The interrupted
treatment gave rise to
drug-resistant forms
of the disease. Free
of TB himself, this
prisoner is allowed
conjugal visits with
his wife but kept
away from disease-
carrying prisoners.

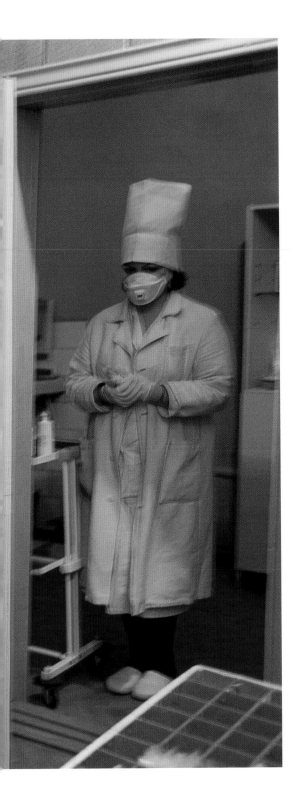

EVEN THE BEST
medicine for TB is no good if it isn't taken regularly, and if the treatment course isn't completed. To ensure that, the new standard, DOT— direct observation treatment—means attendants watch as patients, in this case Siberian prisoners, wash down their pills. Each day inmates cough spittle into a cup to test for the severity of their infections, while doctors check x-rays for evidence of emergent drug-resistant types.

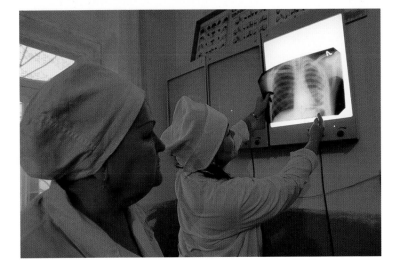

REFUGEES FROM ETHNIC CONFLICTS in Africa crowd into camps, like this one in Tanzania. The nursing woman fled fighting in Burundi with her two children. In crowded camps, bad food, unclean water, and poor sanitation allow illness to thrive.

AFTER THE HORROR
of displacement,
violence, disease, and
despair, a refugee
from the civil war in
the Congo gets ready
to cross Lake Tangan-
yika and begin life
again in her homeland.

ENVIRONMENTAL DISRUPTION

Our ever growing human numbers are placing unprecedented demands on the world's resources and environment. Rivers are dammed to provide more water, forests felled for more timber, or fields cleared for more cultivation. The burning of fossil fuels for energy has begun to alter the atmosphere and, with it, global climate patterns. These threats to the health of the world's environment have a sweeping effect on our own health.

Examples abound of diseases emerging in the aftermath of alterations to a region's ecology. Following the construction of the Diama dam in Senegal in 1986, for instance, epidemiologists saw the emergence and spread of intestinal schistosomiasis, a parasitic disease transmitted by water snails. In Brazil, the destruction of the tropical rain forest triggered huge increases in malaria-carrying mosquito populations, with devastating effects on whole communities of Yanomami Indians.

Global warming may have far more widespread consequences. Mosquito-borne infections such as malaria, dengue fever, and several kinds of encephalitis may already be gaining ground as average annual temperatures rise, expanding the range of the insects. Even in cool highland areas, traditionally free of mosquitoes, insects carrying dengue and yellow fever have been found.

Waterborne diseases may also pose new threats. Reviewing data on climate and disease going back to 1893, researchers recently reported a strong statistical link between rising global temperature, increasingly strong El Niño events, and new outbreaks of cholera in Bangladesh.

Many experts fear that global warming will result in more extreme weather worldwide—longer droughts, fiercer winds, heavier rainfall. Such extremes can give pathogens new avenues for infection. In 1993, a period of drought followed by intense rains in the southwestern U.S. is thought to have ignited the first human outbreaks of hantavirus. Dry weather reduced the population of animals that prey on mice, carriers of the disease, then sudden heavy rains provided abundant food in the form of grasshoppers and pine nuts for the rodents. Their populations exploded, creating a huge reservoir for hantaviruses. When drought returned and the mice were forced to find food and water close to where people lived, the disease spread to humans.

Our health, such outbreaks remind us, is intricately tied to the environment. We share the world, after all, with creatures large and small, including the microorganisms that cause disease. As the pressure of growing human populations leads to the disruption of local and global ecosystems, we're likely to see the consequences impacting the overall health of the planet, ours included.

ARCTIC OIL EXPLORATION on Alaska's North Slope has long inspired political and environmental debate. Larger implications in the debate include questions of how the world's resources should be shared among the world's people.

HAULING IRON ORE,
a 240-car Australian
freight train heads for
port, about half its
cargo earmarked for
Japan. Industrial man-
ufacturing releases
carbon dioxide as a
byproduct, as does
primitive slash-and-
burn farming. The
resultant global
warming may alter
growing seasons and
provide new havens
for dangerous path-
ogens to emerge.

MOUNTAINS DENUDED by strip mining and lungs damaged from breathing the foul air of mines are dark legacies of coal extraction in Appalachia. The need for natural resources and the health costs of their extraction is always a delicate balancing act.

RODENTS HAVE have long been vectors for disease. Data from a New Mexico deer mouse trapped in a plastic bag (left) will help scientists understand the hantavirus, transmitted to humans in rodent saliva, droppings, or urine. Preventive measures include guarding against rodent infestation (right), particularly in the American Southwest, where the most cases are reported.

WITH THE remnants of a failed crop and an abandoned medical clinic as a backdrop, a young boy in Niger is poised precariously between famine and survival. Climate change has increased desertification here, adding to the problems of war, economic dislocation, political corruption, and the loss of manpower to AIDS.

TREES LOGGED IN a Borneo forest are towed downstream, most of them headed for hungry timber markets in the developed world. Throughout many parts of Southeast Asia, Africa, and Central and South America, forests are being razed at rapacious rates. Although the wood itself is prized, large swaths are also being removed to establish farms and cattle pastures. These replaceable ecosystems, now destroyed, may once have sheltered flora and fauna with medicinal potential.

AMONG THE LAST INDIGENOUS peoples to follow their age-old ways, the Penan of Borneo are threatened by a dam project and industrial logging that is reducing their rain forest home at twice the rate Amazonian forests are being felled. The disruption of such traditional lifestyles often leads to chronic physical and mental health problems.

SLASH-AND-BURN clearing prepares new land for villagers of Tierra Blanca (White Earth) in Guatemala's Petén. Ashes from the burned wood will fertilize the soil enough for planting, but because tropical soils are low in nutrients, fertility will soon vanish and new forest will be cut and fired. Along with environmental degradation, the whole process exposes humans to diseases from forest-dwelling microbes.

PARASITES INFECT millions worldwide. From head lice to tapeworms, sleeping sickness to river blindness, deadly pests attach themselves to human bodies and organs. Some of the effects of these pests are actually visible, as on the back of a Guatemalan farmer bitten by a sand fly that carries the leishmania parasite. If untreated, the lesions can leave permanent scars. Such conditions used to affect mostly men working in the forests, but as more people move into formerly wild habitats, women and children, too, are infected.

YELLOW FEVER, though preventable by safe and effective vaccination, still strikes about 200,000 people each year, killing some 30,000 in Africa and the Americas. In Brazil, during wild fava bean season, harvesters who go into the forest to collect the legume are at risk. Medical teams follow them, providing vaccinations and collecting mosquitoes that transmit the disease, as well as human blood samples. Since many victims of yellow fever exhibit only mild symptoms, the disease is transmitted unwittingly, resulting in continued high infection rates.

ANXIOUS BUT
willing, a Brazilian boy
is inoculated against
the yellow fever virus
spread by mosqui-
toes. Viruses cause
diseases ranging from
the common cold to
measles, hepatitis,
and HIV/AIDS. They
infect all living things
but are themselves
not quite alive and
cannot reproduce on
their own. They need
a host—such as a
human. Nobel Prize
winner Peter Medawar
called them "a piece
of bad news wrapped
in a protein."

HIV/AIDS

When the first reports of a rare and deadly form of pneumonia appeared in a small medical journal in 1981, no one could have guessed the scope of the pandemic to come. Even today, almost a quarter century later, it's still difficult to predict the future of HIV/AIDS—except to know that it will continue to be one of the most critical challenges facing global public health.

Over the next decade the number of people infected with the virus is likely to double. Already the losses are almost incomprehensible—young adults who should be at their most productive are sick and dying, millions of children have been orphaned, already fragile health care systems are overwhelmed, and in some parts of the world life expectancy has fallen by decades. Most of this burden is shouldered by developing nations, where 95 percent of new infections occur. In South Africa the prevalence of HIV infection in pregnant women has soared from less than 1 percent in 1990 to 25 percent today. The same catastrophe is being played out in India, China, and countries of the former U.S.S.R. and of the Caribbean, which have the fastest growing rates of new infections.

Beyond its direct toll, the HIV/AIDS epidemic has allowed other infections to flourish, especially those that prey on people with weakened immune systems. Most worrisome is tuberculosis. Once largely controlled, the disease has roared back, killing an estimated two million people a year. It is the leading cause of death among those with HIV/AIDS.

HIV/AIDS has taught us that the threat of infectious diseases is far from over. It has forced us to face up to something else as well: the vast gap between rich and poor. No other disease shows so harshly the disparities between the world's haves and have-nots. In the developed world, widespread access to a host of antiviral drugs has dramatically improved the lives of those infected, turning HIV/AIDS from a killer into a chronic disease. In the poorest nations, people often fall sick and die without any medical care at all. Public health campaigns in places like the U.S. and Europe regularly remind newly sexually active young people of the dangers of HIV/AIDS. Yet in Central Asia and other places where HIV poses an increasing danger, surveys find that as many as nine out of ten older adolescents have never heard of HIV/AIDS.

The HIV/AIDS epidemic can't be ended tomorrow, or next year, or ten years from now. But studies estimate that, using already well-established strategies, at least 28 million infections could be prevented over the next decade. As a 2003 report commissioned by the Global HIV Prevention Working Group concluded, "If we fail to mobilize against AIDS, we will deserve the certain condemnation of history."

A VIGIL FOR AIDS victims is quietly observed in
Washington, D.C. Since the virus was first identified
in the early 1980s, it has claimed millions of lives—
and the count relentlessly continues.

SLAUGHTERED monkey will probably be sold as bush meat in Gabon. Researchers believe HIV/AIDS originated in African primates, then jumped to humans. Tropical forests may contain other reservoirs of potentially lethal viral agents that could be transmitted to humans.

THOUGH SHOOTING UP, drug users have been taught by health organizations to use clean needles to guard against hepatitis and HIV/AIDS. In Russia (above) harsh times have driven many to drug use. A program sponsored by Doctors Without Borders offers vitamins, nutrition advice, and sterile needles.

IN THE RED-LIGHT district of Mumbai (Bombay), AIDS is a particularly acute problem, since many "clients" are truck drivers, who carry their infection with them from place to place, spreading the disease over long distances.

RED-LIGHT LIFE: Mumbai sex workers ready for the evening's business. The children of prostitutes live with their mothers in the city's brothels, where all the women participate in their rearing. In the 1990s the AIDS rate among women here approached 90 percent. Some success in reducing that percentage has been achieved by health groups persuading women to insist that clients use condoms.

PLYING AN ANCIENT craft, a transvestite hooker in Puerto Rico displays "her" wares to potential customers. Prostitution is an increasingly deadly business in these days of the HIV/AIDS pandemic. The rate of AIDS infection worldwide is staggering. At the end of 2002 an estimated 42 million people were living with AIDS, 3.2 million of them children under 15. Prevalence rates were most severe in sub-Saharan Africa—38.8 percent in Botswana —but the virus crosses all borders, races, and ages.

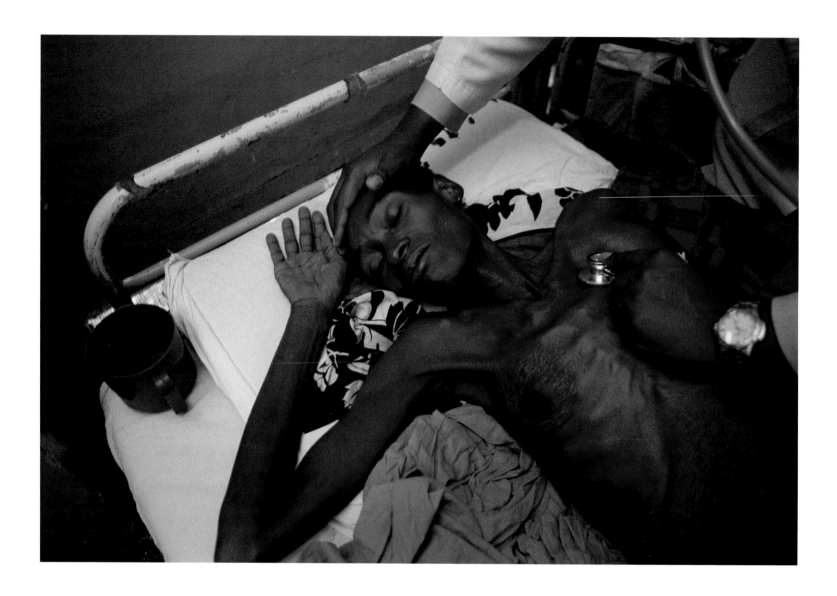

AIDS VICTIMS, like the dying woman opposite, crowd health facilities throughout sub-Saharan Africa; she is in a hospital intended for TB and malaria sufferers. As the death toll rises, graves (above) are a constant and grim reminder of the pandemic.

BATH TIME in a Cambodian orphanage initially established for orphans of the country's protracted civil war now sees the children of AIDS victims scrubbing up. Sadly, because so many of them were infected with HIV/AIDS, the orphanage placed a moratorium on their adoption.

GRIM EXHIBIT (right) is meant to impress Thai military recruits with the consequences of AIDS. Dying victims donated their bodies to this Buddhist hospice to educate others to the dangers of the disease. Above: Cambodian police trainees are taught to put on condoms blindfolded. Cambodia has the highest AIDS rate in Asia, but many sex workers now insist on condom use.

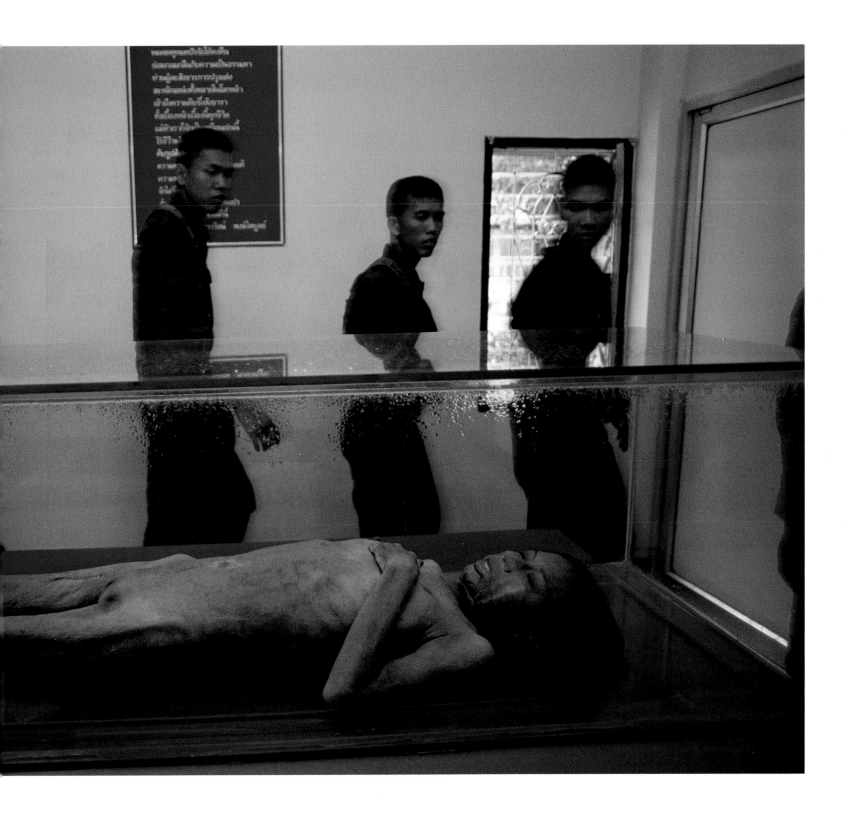

151

INFLATED TO demonstrate its strength, a condom is placed over a patron's head in a San Francisco sex shop. Opposite: Gay men are blindfolded for a class that teaches how affection and physical contact can be shared without the potentially dangerous exchange of bodily fluids.

DYING OF AIDS
alone in a San
Francicso boarding
house, a 50-year-old
man shows a photo-
graph of a daughter
he had not seen since
she was a small child.

AIDS: ONE WOMAN'S STORY

Above all, Fanny loved her children. But when she was pregnant with her second child, she tested positive for AIDS, a development she called "devastating." Her son was born HIV positive but was free of the disease by the time he was 18 months old—long enough for Fanny's antibodies to clear his system. Hearing that news, Fanny declared, "was one of the happiest moments of my life."

For parents, AIDS is not only about them; it's about their children. Children born to HIV-positive women, if fortunate enough to escape the virus themselves, often suffer the loss of one or both of their parents. AIDS has made an indelible mark on the world: There are millions of AIDS orphans to prove it.

Fanny turned her own personal tragedy into a hero's tale, with the help of her physician, Harold P. Katner. Katner saw in her the average middle-class mother, wife, sister, and daughter—someone to whom many people could relate. Believing that Fanny could reach people in ways that a medical doctor could not, Katner persuaded her to join him in a lecturing partnership, and the two began speaking together at churches, schools, and other public meeting places. Fanny told her own cautionary tale with warmth, compassion, and courage: how she dropped out of college, moved in with a bisexual drug addict she later married, and began to use drugs herself. "I fell in love with the coke, with the needle," she says. Her past, she knew, cost her her future.

Even as she lectured, Fanny understood that she was dying, and she became focused on finding a loving home for her two children after she was gone. She admitted that the thought of her children calling someone else "Mommy" was almost more than she could bear. Toward the end, her kids were all that kept her going. "Heaven sounds wonderful to me. I'm ready to go," she said. "The most important thing that keeps me back is the babies."

Fanny died in 1995, just before the various AIDS cocktails that now keep the disease under control became readily available. But she succeeded in her final goal. Her children today are flourishing in a comfortable home, under the loving care of adoptive parents.

LIVING WITH AIDS, Fanny understands that her daughter will grow up without her. Despite, and because of, that harrowing knowledge, Fanny broadcast her mistakes to others, hoping in a small way to fend off the spread of the pandemic.

TWENTY PILLS a day fueled Fanny's regimen of treatment, prescribed to slow the progression of her illness. Her battle was complicated by having to cope with a drug-using husband (left), while raising children. But the quiet moments with her kids—bedtime reading, after-school chats—were immeasurably precious.

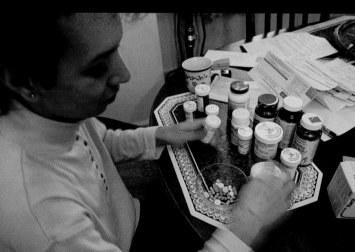

SHARING HER STORY with high school students (opposite) motivated Fanny as well as the students, and earned her hugs. "They can identify with me. I tell them we're all responsible for our own actions." A Christian, Fanny took comfort from her faith.

IN THE AFTERMATH of a family argument, Fanny and her husband wait for the tension to pass. It never really did. His chronic drug problems left the family with one more enemy to battle. "You can decide which way you're going to go, but you're not going to drag down our kids, too," Fanny told him. He eventually died of a brain aneurysm in 1993, leaving Fanny to battle on alone.

III. MEETING THE CHALLENGE

CHAPTER III

MEETING THE CHALLENGE

IN A SMALL VILLAGE IN AFGHANISTAN, HEALTH WORKERS USE LOUDSPEAKERS to call residents to the local mosque—not for prayer but to bring their children to be vaccinated against measles.

In Mali, small teams of international and national health workers and local volunteers travel by bicycles and camels, motorboats and dugout canoes, determined to find and vaccinate every child against polio or to teach people how to protect themselves against Guinea worm disease.

Half a world away, volunteers bring medicine into the slums of Haiti to treat a disfiguring parasitic infection called lymphatic filariasis—drugs that could eventually eliminate the infection by breaking the cycle of transmission.

Public health is practiced on so wide a canvas and in so many different ways—from lifesaving vaccines for infants to water-aerobics classes for senior citizens—that it's all but impossible to make a comprehensive assessment. Still, many signs point to a new and reinvigorated effort to improve the world's health.

One sign is a change in attitude. The global spread of HIV/AIDS, West Nile virus, and now SARS has taught us that disease agents know no borders. Only at our peril do we ignore the conditions that breed new infections—poverty, overcrowding, lack of clean water, and inadequate health care. As long as such conditions exist in the world, all of us, rich and poor, are at risk. In 1999, the Central Intelligence Agency acknowledged that reality, for the first time labeling global disease as a security threat. In 2000, the World Health Organization met to discuss the dangers that AIDS poses to international security—the first meeting of its kind.

Another change is the growing awareness that global economic advancement in even the richest countries is compromised when parts of the world are allowed to remain weakened by illness. Again, HIV/AIDS has offered the most compelling lesson. Besides its appalling cost in human lives, the epidemic has decimated the economies of the hardest hit nations. A disease that claims 80 percent of its victims between the ages of 20 and 50 years strikes at the very people who should be contributing most to a nation's economic progress. And when people in their most

To ward off disease-carrying insects, a family in
Niger sleeps under a bed net. PRECEDING PAGES:
Vaulting into his mid-80s, a California man clearly
enjoys good health—probably the result of consis-
tent excercise, a good diet, and good genes.

productive years die, societies are left with the young and the old, and very few healthy adults to care for them. "In today's increasingly integrated world, if these nations are allowed to disintegrate," former U.S. President William J. Clinton recently warned a gathering of physicians, "the economic, political, and strategic consequences will be felt well beyond their borders."

Indeed, largely because the world is so integrated and interconnected—because it's harder to turn a blind eye to the vast disparities in health that have been allowed to exist—a growing list of organizations have taken up the cause of public health: Lions Clubs International, The Carter Center, Rotary International, the Bill & Melinda Gates Foundation, the United Nations Foundation, and many others. Adding to the resources of long-established organizations such as the U.S. Centers for Disease Control and Prevention, USAID, the WHO, and UNICEF, the commitment of these new partners shows signs of creating a "tipping point," a profound shift from doubt to conviction, skepticism to optimism. The support of high-profile philanthropic groups has drawn the attention of others with money or volunteer resources, convincing more and more people that an investment in the world's health can have powerful and lasting returns.

THE MOST AMBITIOUS PROGRAM UNDER WAY IS THE GLOBAL POLIO Eradication Initiative, begun in 1988 and now nearing its goal of wiping out a disease that once crippled or killed millions of children every year. The scale of the effort is almost unprecedented. In 2000, 550 million children were inoculated—one-tenth of the world's population and 85 percent of its youth. In a single week in 2001, 152 million children in India received polio vaccinations. Cease-fires in conflict-torn regions have been organized to allow vaccinators to reach children at risk. Since 1988, cases of polio have been slashed by 99.8 percent. Poliovirus, which once ran wild in over 125 countries, has been corralled into just 7 countries, where it is now systematically being tracked down and destroyed.

There have been setbacks. The original goal, to eradicate polio by 2000, had to be abandoned when conflicts and logistical problems made it impossible to reach all affected areas. In 2002, an epidemic of the disease flared up in the Indian state of Uttar Pradesh—the largest outbreak in the world since the eradication program began in 1988. Undaunted, the initiative launched a massive immunization campaign in the affected area, with over 80 million children vaccinated in just six days. Experts remain confident that polio will be eradicated by the new target date of 2005. The result: five million children who would otherwise have been paralyzed will be able to walk and run.

The eradication of polio will do more than vanquish this age-old crippler. The campaign has already taken advantage of its new reach to address other health problems. In 60 countries where immunizations were given, vaccinators also gave vitamin A supplements to children. This simple intervention, which dramatically decreases the risk of diarrheal diseases and measles, has prevented an estimated one million deaths. In addition, an extensive international network of laboratories has been established to conduct polio surveillance in order to identify quickly any new cases that occur. With these technologies and expertise in place, the same network can be used to perform other surveillance in the future.

Just as important, the polio campaign has renewed the world's confidence that battles against disease can be fought and won. It has demonstrated that, with enough determination, people in every corner of the globe, from the most crowded cities to the remotest rain forests, can be reached with lifesaving vaccines and medications. If every child can be protected against polio, why not measles? Why not diphtheria, pertussis, and tetanus?

We're still a long way from that goal. More than two and a half million children die every year of diseases that can be prevented with inoculations; experts say about one million deaths could be prevented with wider use of the hepatitis B vaccine alone. In 2002, with support from the Global Alliance for Vaccines and Immunization, China began a 75-million-dollar project to vaccinate infants in its poorest areas. In 2001, the World Health Organization and the United Nations Children's Fund announced a goal of cutting the number of measles deaths in half by 2005. As of this writing, no new cases of measles have been detected in the Americas in half a year, raising hopes that the disease has been eliminated as a health threat in the Western Hemisphere—though in Africa it remains one of the top five causes of death in young children.

OTHER LESS WELL-KNOWN BUT EQUALLY DESTRUCTIVE DISEASES ARE BEING targeted. One is onchocerciasis, or river blindness. The parasite that causes this brutal illness is spread by black flies, which breed in fast-flowing rivers and streams. The bite of a fly releases millions of microscopic worms that cause debilitating itching and, when they invade the eye, blindness. Before campaigns against river blindness began, as many as 10 percent of people in parts of West Africa were completely blind as a result of the disease, and 30 percent suffered serious vision problems.

In the 1980s pharmaceutical researchers discovered a drug that killed the microscopic worms, preventing blindness and helping reduce the spread of the disease. Experts now believe that river blindness can be controlled in Africa and eliminated entirely in the Americas if 85 percent of the population living in endemic areas was treated twice a year. The work is already under way. The drug's manufacturer, Merck & Co., Inc., has agreed to donate it for as long as it is needed. In 1996 The Carter Center, at the request of the founders of the River Blindness Foundation, led a coalition of the WHO, Lions Clubs International Foundation, the World Bank, and a host of nongovernmental organizations to train villagers and health workers in affected areas to administer the drug. By 2007 river blindness may no longer be a public health threat.

Another disabling plague now under control is lymphatic filariasis, a parasitic infection that afflicts 120 million people in at least 80 countries. The threadlike parasite, transmitted by mosquitoes, invades the lymph system, causing grotesque swelling of limbs and other parts of the body. If everyone at risk could be treated once a year for four to six years with drugs that kill the parasite, the cycle of infection between mosquitoes and humans could be broken and the disease eliminated.

A vast effort has begun, led by partners in the Global Alliance to Eliminate Lymphatic Filariasis. GlaxoSmithKline is making albendazole available worldwide at no cost; Merck & Co., Inc., is donating preventive drugs for the regions in Africa where river blindness and lymphatic filariasis coexist. In all, 1.2 billion people must be treated at least five times through a combined drug regimen—an extraordinary effort that will require an estimated six billion tablets.

Meanwhile, a very different kind of eradication effort is ridding the world of dracunculiasis, also known as Guinea worm disease. The life cycle of the waterborne parasite that causes it could have inspired a gruesome horror film. Ingested in contaminated water, Guinea worm larvae grow inside their human hosts, becoming adult worms as long as three feet. After the worms mature, they migrate under the skin, where a painful blister appears. To ease the pain, people often immerse themselves in water, which causes the blister to erupt and the adult worm to emerge, releasing millions of larvae into the water, thus spreading the disease. Today, the global burden of this disease has been reduced by 98 percent, from 3.5 million cases in 1986 to about 54,600 in 2002, thanks to an effort led by The Carter Center.

Remarkably, the Guinea worm campaign hasn't relied on high tech vaccines or medications. Instead, grassroots volunteers teach communities to keep infected people out of water sources and to filter water through a cloth that removes worm larvae. Such simple but effective interventions have a long and distinguished history in public health—ever since John Snow removed the Broad Street pump handle. They still make an enormous contribution. There's nothing high-tech about antismoking public service campaigns or legislation to discourage smoking, for instance. Yet these simple tools have dramatically reduced the rate of smoking in the U.S. and prevented millions of unnecessary illnesses and deaths. Antismoking campaigns under way in many parts of Europe promise to spare many more lives.

Nothing could be more basic than an automobile seat belt. Yet this little device, coupled with safe-driving campaigns, has dramatically reduced traffic deaths and injuries throughout the developed world.

In the poorest corners of the globe, the simpler and less expensive an intervention, the better its odds of making a real difference. Researchers at the Johns Hopkins School of Medicine may have found just such an approach to prevent cervical cancer. Pap smear testing has dramatically reduced deaths from cervical cancer in the developed world, but the test is too expensive and complicated to work in poorer countries, where cervical cancer kills some 225,000 women a year. In a study involving 6,000 women in rural Thailand, experts tested a simple procedure that uses vinegar to wash the cervix, a flashlight to examine it for precancerous lesions, and liquid carbon dioxide to freeze off suspicious spots. Of the women treated, 94 percent had no lesions a year later.

EVEN SIMPLE BARRIERS TO DISEASE ARE LIFESAVERS. CONDOMS REMAIN THE SINGLE most effective AIDS prevention technology around. Bed netting still offers many of the world's poorest people the best defense against malaria. A recent study showed that children who slept under mosquito netting that was impregnated with insecticide were half as likely as unprotected children to develop symptoms of the disease.

A very different kind of barrier is being used to prevent waterborne diseases in Bangladesh: folding clean, dry sari cloth several times and placing it over the mouth of a jug before collecting water. This simple technique filters out a type of plankton called copepod, which carries cholera bacteria. In a study published in 2003, researchers from the University of Maryland chose 65 Bangladeshi villages where cholera occurs. In 27, women were asked to filter water with sari cloth folded eight times. The risk of being infected with cholera was cut in half. Cloth filters also block Guinea worm and may help reduce transmission of other waterborne pathogens.

To be sure, high-tech advances are also transforming the practice of public health. Thanks to global communications systems, the world's surveillance of disease threats is more extensive and responsive than ever before in history. In the aftermath of the spread of West Nile virus into the Western Hemisphere, a new Internet-based surveillance system was established, with the voluntary help of zoos around the U.S., to alert authorities at the first sign of unusual illnesses in zoo populations and wild animals—potential harbingers of human epidemics. Even personal computers have been enlisted in the high-tech battle against disease. When an outbreak of an unusual form of conjunctivitis hit a college campus recently, health workers used campus "blitzmail" to contact students and conduct epidemiological investigations with unprecedented speed. E-mail also enabled them to identify a similar outbreak at another campus.

Still, glaring gaps in surveillance remain. Outbreaks of serious epidemics in the poorest countries in Africa sometimes go unreported, for instance. And SARS is believed to have gained a deadly foothold when officials in China were slow to report the outbreak of an unusual form of pneumonia.

Yet even then, the international health community responded with extraordinary speed once the threat was revealed. Day by day, sometimes hour by hour, officials relayed crucial information about the virus and how to limit its spread via a global network. Researchers quickly confirmed that it was a novel form of coronavirus. Within a few months of the first reported outbreak, genetic sequences of the virus were published. Work began almost immediately to find treatments and develop a vaccine. Never in history has a new infectious agent been identified and dissected so quickly.

In another encouraging example of international cooperation that saves thousands of lives a year, scientists from around the world collect samples of emerging influenza viruses, analyze their genetic makeup, and decide which strains are most likely to go global during the next flu season. Then, each year, vaccine manufacturers fashion a new shot against influenza that matches the most likely emerging strain.

In the years ahead, the single biggest test of global public health is likely to be the HIV/AIDS epidemic. The cost of failure is clear: at least 100 million more people around the world infected with the virus over the next decade, many of them in the world's poorest nations.

Those infections don't have to occur. They are not preordained. Uganda has been a model for what strong leadership and a concerted national effort in a developing nation can do to control the disease. The country has combined public awareness campaigns, extensive condom promotion, and free voluntary testing and counseling with a range of other programs, including many run by people living with HIV/AIDS. In one example, women with the disease have been recruited to distribute containers designed to prevent water stored in homes from being contaminated with waterborne infectious agents. The implicit message is clear and powerful: The battle against HIV is just part of a larger countrywide effort to improve health. And it's working. National HIV prevalence has fallen nearly in half in Uganda since the beginning of the 1990s. The percentage of HIV-positive pregnant women has declined by nearly two-thirds.

Inspiration also comes from Thailand. When infection rates began to soar in the 1980s, the country began a comprehensive public awareness campaign along with a program that mandated 100 percent condom-use programs in brothels. Free condoms were distributed. Police inspectors posing as clients made sure they were being used. Brothels that didn't comply were shut down. By the end of the 1990s, new infections had fallen by 80 percent.

Other countries have also taken important steps in curbing the epidemic. In Côte D'Ivoire, for instance, a vigorous education program increased condom use among sex workers from 20 percent to nearly 80 percent in six years. As a consequence, HIV prevalence fell by almost two-thirds in the targeted population. In Brazil, a combination of AIDS-prevention campaigns, universal access to anti-retroviral drugs, and counseling and support for HIV-positive people has dramatically reduced new infections. In one year, condom use among injected-drug users, one of the most difficult-to-reach populations, increased from 42 percent to 65 percent.

As encouraging as these programs are, they represent only a first step. To turn them into more permanent victories against HIV/AIDS, the international community will need to accomplish an even more difficult task. In the world's poorest places, health care is often fragmentary and sporadic, taking the form of a patchwork of pilot programs, supported by foreign aid. These small programs are as important as any efforts being made in public health. They are saving lives every day. But when they end, the good they accomplish often ends with them. To make a real and lasting difference—not just in the battle against HIV/AIDS but against every preventable threat to human health—the world's neediest countries must be helped to establish sustainable health delivery infrastructures of their own. There's no bigger challenge. A reliable health system depends on political and social stability, on a working economy, on an infrastructure for travel and transportation—and on people who have a real stake in their own futures.

The first important steps toward that goal are being taken. An international consortium that includes donor organizations, health professionals, and business leaders recently launched an initiative designed to help poor countries develop stable and sustainable infrastructures to deliver HIV/AIDS drugs on their own. Efforts are also under way to strengthen the infrastructure created by the polio, Guinea worm, and river blindness campaigns for other uses. In India and Africa, banks have begun making microloans to women in some of the world's most impoverished places—loans that help them create their own businesses, in turn giving them the cash they need to buy things like safe water-storage containers and mosquito netting.

Bank loans and business plans may seem far removed from public health. But when a mother is empowered to provide her family with clean drinking water, she's more likely to demand that they also get vaccinated against measles. When young people have hope for a better life through education, they have reason to protect themselves against HIV infection. When a family is spared the debilitating cycle of sickness caused by chronic malaria infections, they can begin to build a better life for themselves. The heart of public health, after all, has always been the conviction that every life matters. Its future will depend on making that a reality.

CLEAN WATER, SAFE FOOD

Nothing is more essential to health than clean water and uncontaminated food. Yet in many parts of the world even these most basic requirements are a luxury few enjoy. An astonishing 1.2 billion people on the planet—one in five—do not have access to a reliable supply of uncontaminated drinking water. Twice that number, almost half the world, lacks basic sanitation services. The result: Hundreds of millions of cases of waterborne and food-related infections that could be prevented occur every year.

Supplying clean water and safe food is likely to become a growing challenge as the Earth's human population swells, especially in the world's megacities, where surging numbers of people living in crowded conditions are overwhelming water and sanitation systems. In 1991, a cholera epidemic exploded in Peru and quickly spread through large parts of South and Central America. The cause: contaminated water. In many large cities, the water coming out of faucets carried cholera bacteria. There are other threats. The loss of the world's forests also has an impact on water resources. In a growing number of countries, a recent report by the United Nations Food and Agriculture Organization warned, freshwater supplies are being jeopardized by watershed degradation.

Even where a source of clean water is available, people in many parts of the developing world must walk long distances to get it and then carry it back to their homes—time and energy that could be better spent on more productive activities. Where basic sanitation is lacking, uncontaminated water drawn from wells can easily become contaminated. In a Malawi refugee camp, for instance, researchers from the U.S. Centers for Disease Control and Prevention found that people drawing clean water from source wells quickly contaminated it with their hands, which carried fecal bacteria. Water stored in open containers also became contaminated by rodents and other household pests.

That doesn't have to happen. Researchers have tested a simple approach that combines an inexpensive method of chlorinating water with specially designed storage containers with spigots, to reduce the chances water will be contaminated while stored. Studies in Latin America and Africa have shown this can reduce the incidence of waterborne infections by half. Methods to disinfect water at the point of use using solar power are also showing promise.

So-called point-of-use water decontamination and hygienic storage methods also empower individuals to address the issue of clean water on their own. When members of a poor family in Zambia, India, or Guatemala discover that they can take matters into their own hands and improve their water, they also take the first of what can become many small steps toward a healthier future.

THE SIMPLE ACT OF WASHING with soap and filtered pond water is a powerful precaution against contracting or spreading diseases for this African girl. But clean water, uncontaminated food, and pest-free shelters are luxuries for many in the world.

ON THE OUTSKIRTS of New Delhi, the poor pick through trash under municipal water lines, yet they have no access to clean water. Flooded by new arrivals, megacities throughout the developing world are hotbeds for the growth and spread of pathogens.

DURING FLOOD SEASON, residents of a Dhaka slum manage to get potable water by furtively tapping into a source. Opposite: A bucket shower on the street is the only way an Indian girl has of keeping clean.

SIMPLE SOLUTION:
By filtering standing water through layers of dry, folded sari cloth, a Bangladeshi woman can screen out the plankton to which cholera bacteria adhere. The water-borne disease once ranked among the most widespread and deadly scourges in the world.

FOREIGN DELICACY: In a 1990s outbreak, Guatemalan raspberries carried the cyclospora parasite to 2,000 Canadians and Americans. Outhouses on the edge of fields and pickers with dirty hands accounted for the illnesses. Growers now spray plants with pesticide and clean water, and workers wash their hands before picking.

DESERT WATER IS especially precious and its sources susceptible to contamination. In Africa, water infested with Guinea worm larvae has long plagued rural peoples. The loathsome affliction now seems to be well on its way to eradication, thanks to a coordinated global effort to educate villagers on preventive measures. Women have been trained to filter water through cloth as they fill the giant gourds used for the long trek back to their villages. And villagers afflicted with the disease have been taught not to enter ponds, where the worms will escape their human hosts and spew new larvae.

A FISH FARM in Norway transports salmon finger-lings from ocean holding tanks to workers who inoculate them against a host of diseases, thus preventing the use of antibiotics later. In Norway, healthy foods are a source of national pride.

BY GOVERNMENT mandate, Norwegian farms are small enough for the farmers to know their livestock personally—a practice that makes for healthier, more stress-free animals. Norway has used its wealth from offshore oil wells to create a self-sufficient and disease-free food supply. Cows who don't live the free-range life are more subject to disease, so a U.S. researcher (left) tests a spray to prevent *E.coli* bacteria from entering beef.

ON A POULTRY farm in Norway, chicks are checked every day for salmonella bacteria. If even one is infected, all must be destroyed.

EMPOWERING WOMEN

From Sierra Leone to Seattle, from London to Kuala Lampur, women are often chiefly responsible for their family's health. In many parts of the world, in fact, women are the sole heads of the household, especially where men have been forced to leave because of war or economic want. Empowering women through economic opportunity can be the first step in improving family health and well-being.

One of the most inspiring success stories comes from Bangladesh, where the Dhaka-based Grameen Bank has been loaning small sums of money since 1976. The loans are designed to help women start small businesses and begin earning money—often for the first time in their lives. Some women sell handmade crafts. Others make bamboo furniture. Others are more innovative. One woman in Bangladesh used her loan to buy a cellular phone and now earns money by selling telephone calls. To date, the Grameen Bank program has 2.4 million borrowers, including women in 41,000 villages that represent more than 60 percent of the total villages in Bangladesh.

Similar programs are helping create income-earners in other poor countries. In Mexico, for instance, 35 institutions now loan money to the country's poorest. One of them, the Mexico City-based Compartamos, which means "Let's Share," has 140,000 clients and a loan portfolio of $45 million. Throughout Mexico, women have borrowed and repaid tens of millions of dollars in the past two years alone. Even a few hundred dollars can offer hope for a better, healthier life. With a loan of about $250, a 31-year-old widow in one of Mexico's poorest areas began making clay pots, which she sells through a wholesaler. She now earns $15 to $20 a week, doubling the family's income.

It's no coincidence that 95 percent to 98 percent of the loans are made to women. Women given an opportunity to earn money, research shows, are more likely than men to spend it on their family's health. One study tallied a 20-fold greater impact on children's health when income is controlled by the mother instead of the father.

Loan programs have done more than help women start small businesses; they've also encouraged them to begin saving money. Those savings represent a degree of stability that many very poor families never had before. Giving women the power to make economic decisions also empowers them to make other decisions, including planning family size—another important contributor to health and well-being.

SCHOOL DAYS for girls in Bangladesh herald better
times for the whole society. Studies show that empowering women improves well-being overall—stronger
economic potential for families, fewer children, and
better health.

CELL PHONE technology allows a Bangladeshi woman to pass on the sad news of her husband's death. The cell phone is also a lifeline to her economic well-being: With $390 borrowed from Grameen Bank, she bought the phone as an investment, renting time on it to other people in her village. The innovative bank requires a pledge for its loans: "We shall plan to keep our families small. We shall keep free from the curse of dowry. We shall not practice child marriage." Loans for small businesses—such as cell-phone rental and dyeing cloth for saris —give women independence and hope.

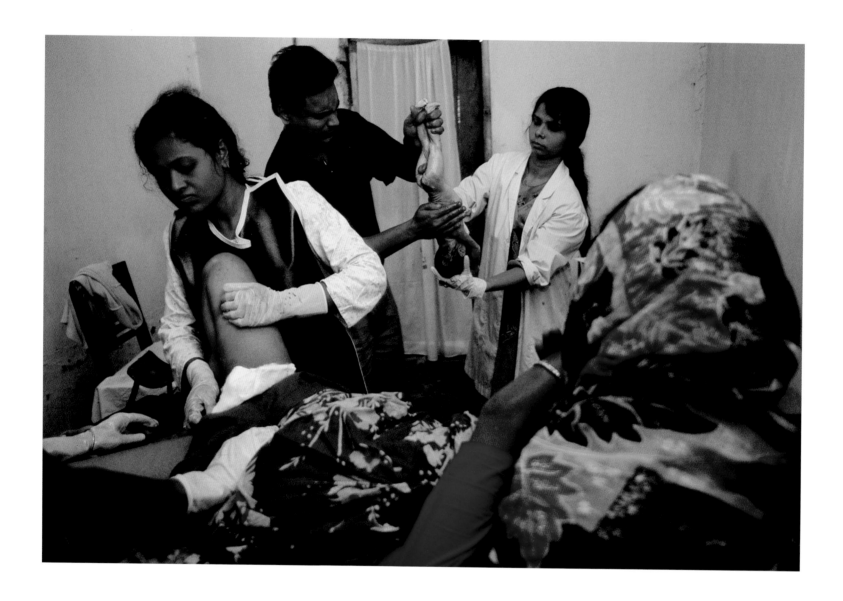

THE BIRTH OF A CHILD by a child is part of the cycle of dependency leading to poverty that many public health organizations are working to change. This Bangladeshi teenager endured labor for three days and was still unconscious a day after giving birth.

HEAD START ON A
better society begins
with African girls
whose parents can
afford to send them to
school. They will likely
grow up with enough
education to post-
pone marriage and
childbirth, to teach
their own children—
the few they bear—
the economic facts of
life, to assert them-
selves in finding good
employment, and to
help provide for their
own families' futures.

WOMAN TO WOMAN

Social agencies and NGOs (nongovernmental organizations) working to better the lives of people long ago realized that, if you empower women, you empower whole communities. The empowerment can come through many avenues: granting microloans, offering educational programs, or training women to be village volunteers, a nonpaying job that carries with it esteem and respect.

In places like Bangladesh, one of the poorest countries in the world, these village volunteers have become the frontline soldiers in the seemingly insurmountable task of delivering reasonable health care to as many people as possible. Without their efforts at the village level, the most extensive health plan would be simply a cerebral exercise.

To earn the privilege of being volunteers, women must be respected members of their communities. They must have good husbands and be good wives. That includes keeping a clean home, treating their children well—and limiting the number of those children.

Because of the esteem and trust they have among their fellow villagers, these volunteers are able to address effectively the health concerns of women and children in their communities. They oversee the dispensing of medicines and follow up to see that the medicines are taken properly. They also act as conduits for the dispersion of information between their villages and organizations wanting to deliver care or supplies.

Educating girls and giving women the ability to earn their own money also empowers these women to reach out to other women in their communities, to help them achieve a better and healthier life for themselves and their families.

Since 1975, Bangladesh has lowered its birthrate from seven children per woman to fewer than four. Though poverty and attendant health problems still plague most people, programs like the village volunteers are paving the way to a better future.

WITH QUIET SUPPORT from a local imam, a volunteer with the Bangladesh Rural Advancement Committee makes her rounds, primarily counseling women on family planning and health care.

DISPENSING ANTIBIOTICS for a mother's sick child, the Bangladesh Rural Advancement Committee volunteer may take the opportunity to discuss contraception and offer instruction on the use of various methods. Later, she completes her own household chores (right).

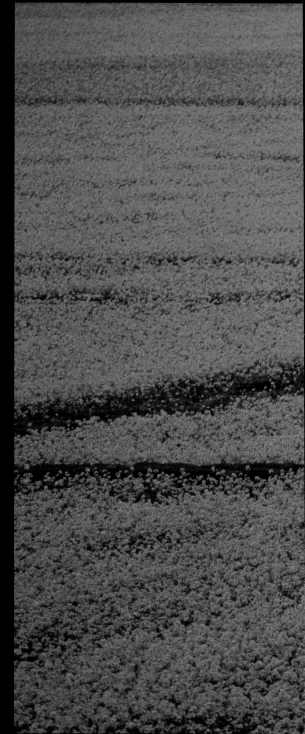

Surrounded by the gold of blooming rape, a
Bangladeshi woman tends the goat that ensures
her a small income of her own. Above, a tailor who
started her business with a small loan takes a quiet
moment away from her busy shop, a meeting place
for local women.

TAKING CONTROL of her future, a Bangladeshi widow borrowed money from Grameen Bank to start her own carpentry business. With help from her children, she also makes furniture and has even managed to build a home for herself and her family.

VACCINES

The power of vaccines is nothing short of miraculous. The world's first vaccine, against smallpox, ultimately eradicated the disease itself. Before long another vaccine—a few drops placed on a child's tongue to prevent polio—will banish that disease from the face of the Earth, as well.

Even when vaccines can't eliminate afflictions, they can offer lifesaving protection. Measles immunization prevents an astonishing 2.7 million deaths a year, according to the CDC. The combination of diphtheria, pertussis, and tetanus vaccines given to infants protects against diseases that were once the scourge of childhood. Yellow fever vaccine has eliminated the threat of that disease from much of the world.

Today, some 250 years after a physician named Edward Jenner discovered that inoculating people against cowpox offered protection against the much deadlier smallpox, researchers are employing the advanced tools of genetic engineering to develop new vaccines against a host of threats. Several promising HIV/AIDS vaccines, as well as one for West Nile virus, are being tested.

The bigger challenge will be bringing the miracle of vaccines to the people most in need. Two billion people worldwide are infected with hepatitis B virus, for instance, which can cause liver disease and liver cancer. An estimated 350 million people are lifelong carriers of the virus, making them able to transmit it to others. A recently developed vaccine for hepatitis B effectively protects against infection—yet it is only beginning to be made available in parts of the world where the ailment is endemic. Yellow fever vaccine was first developed in the 1930s. It remains one of the safest and most effective inoculations known. Yet potentially catastrophic outbreaks still threaten. In 2000, an outbreak of yellow fever in the West African country of Guinea threatened the lives of more than a million people. A year later, the disease erupted in Abidjan, the densely crowded capital of Côte d'Ivoire, where 3.5 million people live. An emergency mass vaccination program immunized almost 3 million people in ten days, averting potential disaster.

A new push is under way today to extend the lifesaving reach of vaccinations, which remain by far the most cost-effective public health intervention available. The most dramatic is the Global Polio Eradication Initiative, which is closing in on the last places where the virus thrives. There are also determined efforts to expand measles, hepatitis B, and other immunizations around the world. Vaccination programs require concerted effort and staying power. History shows that when they falter, preventable diseases surge back. But with a determined effort, millions upon millions of lives can be spared.

COAXED BY CANDY, children in India crowd around medical personnel who administer polio vaccines on National Immunization Day. The most difficult task is reaching the last few in the total population. Only then can polio be stopped.

HEALTH WORKERS bring polio vaccines to the people, where they live. Due to the hot climate on the Indian subcontinent, the vaccines must be carried in "cold boxes." In Bangladesh (above) the boxes are a distinctive orange. In India Hindus carry aluminum boxes into a Muslim neighborhood. Some Muslims are suspicious that the inoculations will make their children sterile—a cultural and political complication in this humanitarian effort.

A PAINFUL REMINDER that all children were once threatened by the crippling complications of polio, an Indian boy has to be helped to school by his big sister. Poorer children affected by polio here often end up as street beggars. A global initiative to eradicate the disease in the seven countries where it is still endemic is under way, and the world could be polio free by 2005. In June 2003, Rotary International contributed more than 88 million dollars to the effort.

ALL HANDS AGAINST MEASLES is the message
Kenyan schoolchildren send, while volunteers in Red
Cross shirts troop door-to-door to announce vaccines have arrived. The once common, sometimes
fatal infection is being targeted worldwide.

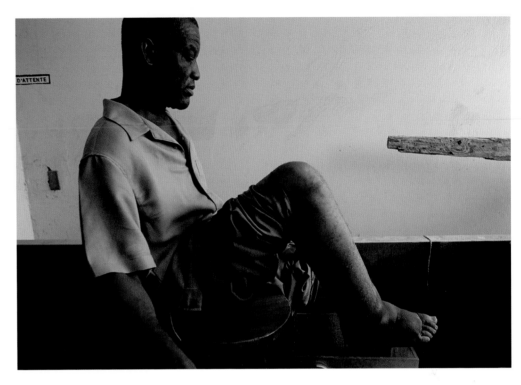

CLEAN, CLEAR WATER is a rare commodity in Haiti. Where water is dirty and stagnant, mosquitoes breed and carry elephantiasis. Above, a sufferer waits for treatment of the disfiguring, lymphatic affliction in a clinic. In severe cases, limbs may have to be amputated.

HANDFUL OF HOPE:
Haitian schoolgirls
know that this mix of
medications, taken
annually, will prevent
lymphatic filariasis—
elephantiasis—
a disease caused by
mosquito-borne
parasites invading the
lymph system.

GENETICS

The beginning of the 21st century marked one of the greatest accomplishments in scientific history—one that is likely to have enormous consequences in coming years: The sequencing of the human genome, the complex of 100,000 genes that spells out the instructions for the creation of a human life.

The impact of this astonishing achievement will resonate in virtually every field of health. Already, scientists have identified genes linked to Alzheimer's disease, breast cancer, colon cancer, cystic fibrosis, Huntington's disease, and a long list of other ailments. At the leading edge of medicine, researchers have begun replacing defective genes, often using benign viruses to carry new genes into cells. The cure for cancer could well lie in genetic research. So could fixes for devastating mental illnesses like schizophrenia and severe depression. In the brave new world of medicine's future, genes that predispose people to a wide array of ailments could be eliminated at the start of life.

That future is years off, to be sure. But the revolution in genetic research already has yielded wonders. Not only the human genome but the genetic codes for many disease agents have been worked out, from smallpox and polio to HIV. The techniques of genetic analysis allow researchers to trace the evolution of emerging viruses and better understand the mechanisms of drug resistance. By studying the way existing genes are reassorted in nature to create new influenza viruses, molecular epidemiologists can spot the emergence of especially dangerous new bugs, giving drugmakers time to create new vaccines. Genetic tools have also revolutionized vaccine development, making it possible for researchers to mix and match pieces of existing viruses with the antigenic fingerprints of targeted organisms. A promising new vaccine against West Nile, for instance, uses the antigen of that virus snipped onto the tried-and-true yellow fever vaccine, one of the safest and most effective immunizations.

Along with its extraordinary promise, however, the revolution in genetics also poses grave ethical dilemmas. It's now possible to test individuals for genes linked to disabling and even deadly diseases for which there is no cure or treatment. What do we do with such awesome knowledge, when there is still nothing that can be done to prevent the disease? Simple genetic screenings will be able to assess an individual's inherited risk of a growing list of illnesses, from osteoporosis and cancer to heart disease and Alzheimer's. How will we ensure that such information is used to help rather than discriminate against people at increased risk? How will we learn to use the extraordinary power we now have to look into an individual's health future? Addressing those questions may prove as difficult, in its way, as the deciphering of the human genome.

BRED FOR HEALTH—human health—these piglets are descended from a pig whose genetic makeup was altered by insertion of a human gene producing a protein needed by hemophiliacs. The protein in their milk will be used in treatment.

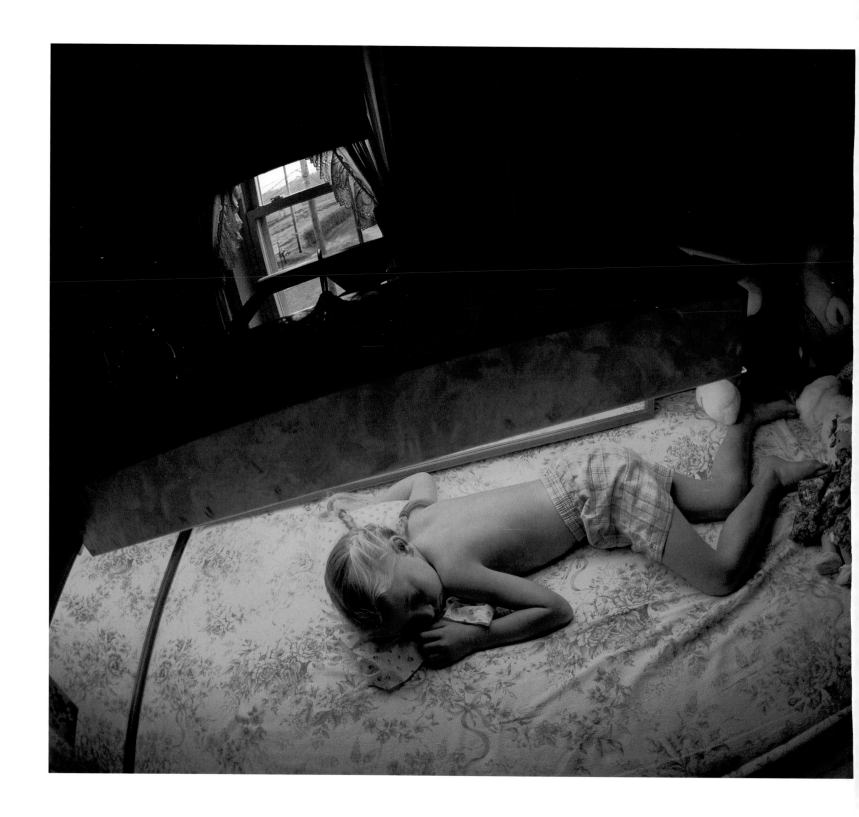

DNA GLOWS RED
in blue light. The double-helix molecule is the basis of all life, but sometimes errors creep into the way it manifests itself through human genes. A Mennonite girl needs to sleep under blue light to help rid her of toxic levels of bilirubin, a blood product that her liver cannot break down. Half of such cases in the U.S. occur in Pennsylvania Amish and Mennonites, who may have inherited the mutation from one founding family 16 generations back. Though gene therapies show great promise, their results are still uneven.

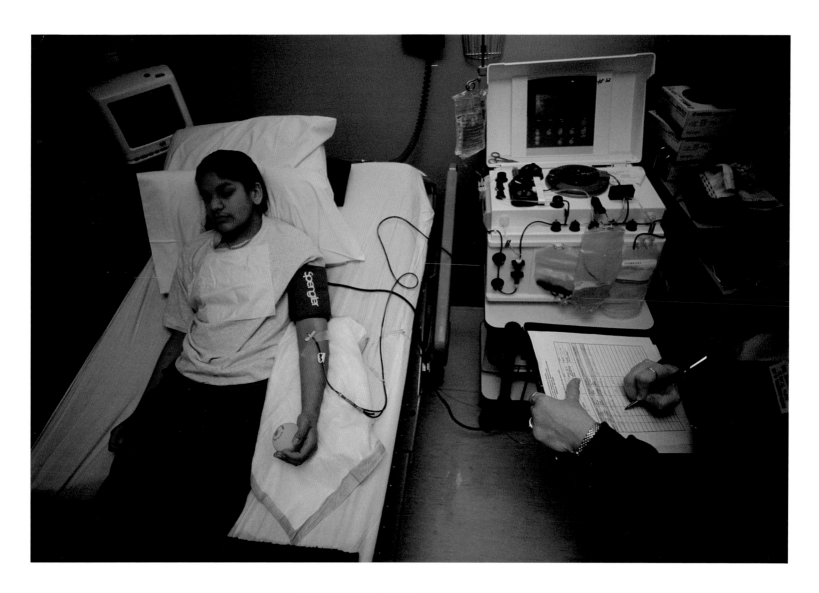

THE FIRST EVER GENE therapy was used
successfully to treat this girl who suffers from
a rare immune disorder (above). High levels of
bilirubin cause chronic jaundice in this Mennonite
girl (opposite).

MATRIX OF TISSUE samples from military personnel, kept on file at a government facility in Maryland, are stored for identification purposes. Such libraries of human essence bother some observers. Is privacy invaded? To what unauthorized ends could the samples be put? The realm of genetic research and its uses raise serious moral, ethical, and political questions not easily answered.

HEALTHY AGING

Good health has created an unexpected challenge in many parts of the world: the graying of the population. People lucky enough to live in the world's most prosperous nations enjoy life spans that would have astonished earlier generations. The Japanese live to an average age of 80. The median life expectancy in the U.S. is 77. In America, only 1 in 25 Americans was 65 or older in 1900. By 1990 that number had increased tenfold, to 1 in 8. In 2000, the United States census form for the first time included a three-digit space for age—acknowledging the 70,000 centenarians alive in the country and marking a profound demographic shift. By 2050 the number of people 65 and older is expected to double from what it is today.

The aging of the human population will shift the burden of disease to chronic illnesses like cancer, heart disease, and diabetes, which typically show up later in life. The cost of caring for growing ranks of older people will add more pressure to already stretched health care systems. Surveys find that the majority of older people have at least one and often several chronic conditions, including arthritis, hypertension, heart disease, cataracts and hearing impairments. People over 65 are more likely to be hospitalized and to spend more time in the hospital than their younger counterparts. They also have twice as many contacts with doctors. With age, the prevalence of severely disabling illnesses like Alzheimer's increases —diseases that place an enormous burden on families and society.

But many of the diseases long associated with age aren't inevitable, we're learning—heart disease, for instance. Researchers at the Harvard School of Public Health, who have been tracking the lifestyles and health of more than 120,000 nurses around the U.S., recently reported that women who followed basic advice to exercise, avoid smoking, and eat a healthful diet cut their risk of cardiovascular disease by more than 80 percent. Other studies show that a simple regimen of physical activity helps older people continue to function independently and lowers their risk of injury from falls—a major health threat in old age. One-third of cancer cases are tied to diet, experts say. New evidence suggests that maintaining a healthy weight can also dramatically reduce the risk of a variety of cancers.

Experts say it's never too late to start making healthful changes. But there's no better time than when people are young. That's an important message not only for countries with graying populations but for those with booming populations of young people. In India, for example, almost half the population is under the age of 15. In Egypt, the median age is a mere 22. Promoting healthful lifestyles in these countries now will have an enormous impact on health in years to come.

HANDSTANDING IN HIS 80S, George Nissan,
lifelong gymnast and inventor of the trampoline,
shows off for his grandson. If the child takes after his
granddad, could he be doing the same in 2082 or
thereabouts?

NO GIDDYAP needed as Grace Boyd (opposite), widow of film cowboy Hopalong Cassidy, teaches a tai chi class. Above, identical Danish twins are part of a study on the relationship of genetic and environmental factors to aging.

MERMAIDS IN training practice synchronized moves for an annual show at their retirement community in California. Healthy lifestyles, medical advances, and optimistic attitudes are keeping many people going strong well past what used to be called "the prime of life." A long, healthful, and productive life is everyone's dream. For more and more people, it's becoming a reality.

HOTLINES IN U.S.

National Immunization Hotline
  800-232-2522 (English)
  800-232-0233 (Spanish)
National AIDS Hotline
  800-342-2437 (English)
  800-344-7432 (Spanish)
National STD Hotline
  800-227-8922
Traveler's Health 877-394-8747

HEALTH ORGANIZATIONS

World Health Organization
(WHO)
The World Health Organization
was established in 1948 by the
UN. Its stated objective is "the
attainment by all peoples of the
highest possible level of health."
http://www.who.int/en/
Avenue Appia 20
1211 Geneva 27
Switzerland
Tel: 41-22-791 21 11
Fax: 41-22-791 3111
Telex: 415 416
Telegraph: UNISANTE GENEVA

Pan American Health
Organization (PAHO)
The Pan American Health
Organization has 100 years'
experience working to improve the
health and living standards of the
people of the Americas. It serves
as the WHO's regional office and
as the health arm of the Inter-
American System.
http://www.paho.org/
Pan American Sanitary Bureau
Regional Office of the WHO
525 23rd St., NW
Washington, DC 20037
U.S.A.
Tel: 1-(202) 974-3000
Fax: 1-(202) 974-3663

Centers for Disease Control and
Prevention (CDC)
The Centers for Disease Control
and Prevention (CDC) is recog-
nized as the lead federal agency
for protecting the health and safety
of Americans—at home and
abroad. It provides credible information to enhance health decisions and promotes health
through strong partnerships.
http://www.cdc.gov/
1600 Clifton Rd.
Atlanta, GA 30333
U.S.A.
Tel: 1-(404) 639-3311

American Public Health
Association (APHA)
APHA has been influencing poli-
cies and setting priorities in public
health for over 125 years.
http://www.apha.org/
800 I St., NW
Washington, DC 20001
U.S.A.
Tel: 1-(202) 777-2742
TTY 1-(202) 777-2500
Fax: 1-(202) 777-2534

National Mental Health
Association (NMHA)
The National Mental Health
Association (NMHA) is the oldest
and largest nonprofit organization
addressing all aspects of mental
health and mental illness in the
U.S. It works to improve the
mental health of all Americans
through advocacy, education,
research, and service.
http://www.nmha.org/
2001 N. Beauregard St., 12th Floor
Alexandria, Virginia 22311
U.S.A.
Tel: 1-(703) 684-7722 or (800)
969-6642
TTY: (800) 433-5959
Fax: 1-(703) 684-5968

Global Health Council (GHC)
The Global Health Council is the
world's largest membership
alliance dedicated to saving lives
by improving health throughout
the world.
http://www.globalhealth.org/
1701 K St., NW
Suite 600
Washington, DC 20006-1503
U.S.A.
Tel: 1-(202) 833-5900
Fax: 1-(202) 833-0075
E-mail: ghc@globalhealth.org

American International Health
Alliance (AIHA)
The mission of the American
International Health Alliance
(AIHA) is to advance global health
through volunteer-driven partner-
ships that mobilize communities
to better address healthcare priori-
ties, while improving productivity
and quality of care.
http://www.aiha.com/
1212 New York Ave., NW
Suite 750
Washington, DC, 20005
U.S.A.
Tel: 1-(202) 789-1136
Fax: 1-(202) 789-1277
E-mail: webmaster@aiha.com

International Health Exchange
(IHE)
International Health Exchange
supports initiatives to bring about
sustained improvements to health
in developing countries by providing experienced people to organizations requiring their skills.
http://www.ihe.org.uk/
1 Great George St.
London SW1P 3AA
England
Tel: 44-0-207 233 1100
Fax: 44-0-207 233 3590
E-mail: info@ihe.org.uk

International AIDS Society (IAS)
The International AIDS Society is
the world's professional society for
scientists, health care and public
health workers, and others
engaged in HIV/AIDS prevention,
control, and care.
http://www.ias.se/
Karolinska Institute
Berzeliusväg 8
171 77 Stockholm
Sweden
Tel: 46-8-508 846 40
Fax: 46-8-508 846 64
E-mail: secretariat@ias.se

Post-Polio Health International
(PHI)
Post-Polio Health International's
mission is to enhance the lives
and independence of polio survivors and home mechanical-
ventilator users through network-
ing and advocacy.
http://www.post-polio.org/
4207 Lindell Blvd., #110
Saint Louis, MI 63108-2915
U.S.A.
Tel: 1-(314)-534-0475
Fax: 1-(314)-534-5070
E-mail: info@post-polio.org

International Union Against
Tuberculosis and Lung Disease
(IUATLD)
The IUATLD is devoted to the prevention and control of tuberculosis
and lung disease and related
health problems, with a particular
emphasis on developing countries.
http://www.iuatld.org/
68 Blvd. Saint Michel
75006 Paris
France
Tel: 33-1-44.32.03.60.
Fax : 33-1-43.29.90.87.

ACKNOWLEDGMENTS

We would like to offer our deep appreciation to GlaxoSmithKline for their gift to publish this book and the support of their staff, especially Dr. Brian Bagnall. We owe special thanks to the Office of Global Health, Centers for Disease Control and Prevention, for its generous assistance, and in particular to Dr. Stephen Blount, Kristen McCall, and Monique Petrofsky, as well as to The Carter Center. We extend warm thanks to Dr. William Foege of the Bill & Melinda Gates Foundation, Pam Wuichet of Project Resource Group, and Dr. Jeffrey P. Koplan at Emory University's Woodruff Health Sciences Center. For their insights, special thanks to Dr. Muireann Brennan and Dr. Daniel Kertesz.

For their ongoing support over the years, Karen Kasmauski would like to thank the following staff and former staff at NATIONAL GEOGRAPHIC: Editor in Chief Bill Allen; Kent Kobersteen, Susan Smith, and Tom Kennedy on the Photographic staff; Illustrations editors Bill Douthitt, David Arnold, and Todd James; Constance Phelps and the Design staff; and Mary McPeak for her research guidance and willingness to lend an ear. In Mission Programs, her appreciation goes to Terry Garcia, Rebecca Martin, and Charlene Valeri. Thanks also to Lisa Lytton, Karen Kostyal, and David Griffin.

Peter Jaret would like to thank Rick Gore, former science editor at NATIONAL GEOGRAPHIC, for his support. Special thanks to Steven Peterson for lending an ear and the occasional encouraging word.

INDEX

A

Africa: disease control 14–15; Guinea worm disease 9; HIV/AIDS 38; measles 170; refugees **4**; reproductive health 10; water **186–187**; women 10
African Program for Onchocerciasis Control (APOC) 15
Aging: research 233; *see also* Elderly people
Agriculture **190–193**; danger to workers 54; pollution 114
AIDS *see* HIV/AIDS
Alaska: oil exploration **113**; oil spill **19**
Alzheimer's disease **74–81**
American Cyanamid/BASF (company) 13
Amish: inherited disorders 225
Animal-borne diseases 52, 84, 86, 88, 112, 118, 137; surveillance 172–173
Anthrax 23, 82
Antibiotics 49–50, 189
Appalachia (region), U.S.: mining **116**
Aral Sea, Kazakhstan-Uzbekistan **101, 102–103**
Automobiles: seat belt use 171

B

Bacteria 49–50
Bangladesh: birthrate 56, 202; childbirth **198–199**; cholera prevention 172, **182–183**; economy 194; education of girls **195**; health care **36–37**; loans to women 194, 196; malnourished children **30–31**; newborn **16–17**; polio vaccinations **11, 212**; pollution **34–35**; tetanus immunization 13; water system 53, **181**; women 194, **196–197, 202–209**
Bangladesh Rural Advancement Committee: volunteers **202–209**
Belém, Brazil: breastfeeding mother with twins **57**

Bill & Melinda Gates Foundation 13, 168
Bioterrorism 23
Birthrates 47, 56
Blindness 12; prevention 9, 13, 14–15, 170, 171
Borneo: indigenous people **124–125**; logging **122–123, 124–125**
Botswana: HIV/AIDS 145; public health initiatives 55
Bovine spongiform encephalopathy 52, 82
Boyd, Grace 232
Brazil: HIV/AIDS prevention 174; malaria 112; vaccinations **132–133**; yellow fever **130–131**
Breastfeeding **57**, 62
Bubonic plague 51
Burundi: warfare 109

C

Calexico, California **94–95**
Cambodia: HIV/AIDS prevention **150–151**; orphanage **148–149**
Canada: mad cow disease 52, 82; SARS 46
Cancer 230; prevention 172; treatment **32–33**
Cardiovascular disease: and lifestyle 28, 54, 230
Carter Center 8, 14–15, 168, 170; Guinea worm campaign 171; International Task Force for Disease Eradication 13
Centers for Disease Control and Prevention, Atlanta, Georgia 82; Epidemic Intelligence Service (EIS) 15
Cervical cancer: prevention 172
Chernobyl, Ukraine 53–54
Childbirth **198–199**; deaths during 10
China: disease surveillance **86–87**; duck farm **84–85**; influenza 50; SARS 50; slaughterhouse **86–87**; vaccinations 169
Cholera 12, 20, 176; prevention 172, **182–183**

Chronic illnesses: and lifestyle 28, 230

Civil wars 4, 100, 111

Climate change 112, 114, 120–121

Clinton, William J. 168

Cold, common 88; research 89

Congo: civil war 111; *Ebola* 53

Coronavirus 48, 88

Côte d'Ivoire: HIV/AIDS prevention 174; yellow fever outbreak 210

Cyclospora parasite 185

D

Deforestation 122–123, 124–125, 176

Dengue fever: prevalence 112; spread 82

Denmark: elderly people 233

Dhaka, Bangladesh 44–45, 181

Diabetes: and obesity 28, 54

Diarrheal diseases 12, 31, 53, 169

Doctors Without Borders 102, 138–139

DPT: vaccine 55, 210

Dracunculiasis *see* Guinea worm disease

Drug-resistant germs 21, 49, 104

Drug use 138–139, 156, 162; and HIV/AIDS 139, 174

Dupont (company) 13

E

E. coli outbreaks 52, 82; prevention 190

Ebola 40, 53

Education 42, 200–201; Kenya 60; United States 64–65

Egypt: median age 230

Elderly people 233; Alzheimer's disease 74; care 26–27, 47; exercise 164–165, 231, 232, 234–235; healthy 230–235; *see also* Aging

Elephantiasis *see* Lymphatic filariasis

Emergent diseases 46, 48–50

Encephalitis 112

Environmental degradation 53–54, 112–133, 114, 176; Alaska 19; Aral Sea, Kazakhstan-Uzbekistan 101, 102–103

Epidemiology 15

Europe: antismoking campaigns 171; immigration 98–99; malaria 112; population 56

European Union: foreign aid 8

Exercise 72–73, 164–165, 231–232, 234–235

Exxon Valdez (oil tanker): oil spill 19

F

Family planning 12, 56, 62; counseling 204

Fertility rates 47, 56

Food-borne pathogens 52, 82, 176, 184–185

Foot-and-mouth disease 52, 82

G

Gambia: acute respiratory infections (ARI) 12

Gates Foundation 13, 168

Genetics 222–229

Germany: SARS 46

GlaxoSmithKline 13, 171

Global Alliance for Vaccines and Immunizations 169

Global Alliance to Eliminate Lymphatic Filariasis 171

Global Polio Eradication Initiative 168–169, 210

Global warming 112, 114, 120–121

Gross national product (GNP) 8

Guangdong (province), China: SARS 46

Guatemala: agriculture 184–185; parasites 128–129; slash-and-burn farming 126–127

Guinea: yellow fever outbreak 210

Guinea worm disease 8–9, 11, 12; eradication 13, 54, 171; prevention 172, 187; public health campaign 171

H

Haiti: clean water 218–219; lymphatic filariasis 219–221; schoolgirls 220–221

Hantaviruses 52, 112; prevention 118–119

Health care 8; workers 14

Heart disease: and lifestyle 28, 54, 230

Hemophilia: treatment 223

Hepatitis: prevention 139; transmission 52

Hepatitis B: vaccine 169, 210

HIV/AIDS 21, 48, 53, 134–163; infection rate 38, 55, 134, 142, 145, 173, 174; and international security 166; origin 137; orphans 148–149; and population growth 56; prevention 134, 139, 142, 150–153, 172–175; public health campaigns 173–175; spread 100, 141; transmission 52; treatment 134, 156, 158; vaccines 210; victims 38–39, 146, 156–163; vigil 135

Ho Chi Minh City, Vietnam 58

Hong Kong: SARS 46

I

Immigrants, illegal 96–97, 98–99

Immunizations *see* Vaccines and vaccinations

India: clean water 181; median age 230; polio outbreaks 169; polio vaccinations 168–169, 211, 212–213; polio victims 214–215; water system 178–179

Indigenous people 124–125

Infant morality 10

Infectious diseases 7, 25, 40; death toll 53; emergence 46, 48–50; spread 51–53, 82–99; surveillance 52–53, 82, 86–87, 172–173

Influenza 46, 53; deaths 84; new strains 48–49, 50, 84, 86, 173; vaccines 173

Insect-borne diseases 9, 112, 130; prevention 167, 170–171, 172

Italy: immigration 98–99; SARS 46

Ivory Coast: HIV/AIDS prevention 174; yellow fever outbreak 210

J

Japan: elderly people 26–27, 47; life expectancy 230

Jenner, Edward 210

Junin virus 51

K

Katner, Harold P. 156

Kenya: family planning 62; Masai children 60–61; measles eradication effort 216–217

L

Leishmania 128–129

Life expectancy 10, 230

Lifestyle 54, 230; impact on health 66, 67–73

Lions Clubs International 13, 168, 170

London, England: cholera 20; sanitation 20

Lyme disease 50–51

Lymphatic filariasis 13, 170–171; eradication 13, 15, 170–171; prevention 220–221; victim 219

M

Mad cow disease 52, 82

Malaria: drug resistance 49; and global warming 112; impact 21–22, 53; prevention 172; spread 82

Malawi: cholera 176

Malnourishment 10, 12, 30–31

Masai people 42–43, 60–61

Measles 12, 53; eradication 13, 14, 22, 169; prevalence 22; vaccinations 13, 14, 169–170, 210, 216–217

Medawar, Peter 132

Meningitis 12

Mennonites: inherited disorders 224–225, 226–227

Merck & Co. Inc. 13, 170, 171

Mexicali, Mexico 94–95

Mexico: border with U.S. 94–95, 96–97; disease surveillance 82; loans to women 194

Mining 116

Monkeypox 52

Mosquito-borne diseases *see* Insect-borne diseases

Mumbai, India: HIV/AIDS 140–143; prostitution 140–143

N

Nebraska: teenage mothers 64–65
Niger 120–121, 167; smallpox victims 93
Nigeria: Guinea worm disease 12; health education 14–15; river blindness 14–15
Nissan, George 231
Norway: agriculture 190–193; fish farm 188–189
Nuclear accidents 53–54
Nutrition: education 31; supplements 169; undernutrition 10, 12, 30–31

O

Obesity 28–29, 54
Oil exploration 113
Onchocerciasis *see* River blindness
Orphans 148–149

P

Parasites 128–129
Pasteur, Louis 7
Penan people 124–125
Peru: cholera outbreak 176
Pfizer (company) 13
Philippines: celebration 2–3; tobacco use 67
Polio: eradication 13, 14, 168–169, 210; vaccine 11, 55, 168–169, 211–213; victims 214–215
Political upheaval 4, 100–111; *see also* War
Pollution 34–35, 53, 101, 114, 187
Population 2–3, 6; density 12, 50, 56; growth 6, 36, 47, 50, 56, 59; and infectious diseases 25
Pregnancy: health care 36–37; and HIV/AIDS 134, 156; teenage 64–65, 198–199; *see also* Childbirth
Primary health care services 14
Prostitution: and HIV/AIDS 140–145, 174
Public health: developing coun-

tries 14–15, 55, 175; funding 13; mission 20
Public health campaigns 171
Puerto Rico: prostitute 144–145

R

Radiation 53–54; heavy-ion 32–33; therapeutic 32–33
Refugees 110–111; camps 108–109, 176
Respiratory illnesses 12, 102, 117
River blindness 12; control 9, 13, 14–15; eradication 13, 170; prevention 170, 171
River Blindness Foundation 170
Rodent-borne diseases 52, 112, 118–119
Rotary International 13, 168, 215
Russia: drug use 139; tuberculosis 53, 104, 104–105, 106–107
Rwanda: civil war 100; waterborne diseases 100

S

Salmonella bacteria 193
San Diego, California: exercise class 72–73; obesity 28–29; tract housing 59; U.S. Navy recruits 68–69
San Francisco, California: HIV/AIDS prevention 152–153; HIV/AIDS victims 154–155
Sanitation 34–35, 109, 176
SARS (Severe acute respiratory syndrome) 46, 48, 51–52, 82, 83, 173
Schistosomiasis 112; control 15
Senegal: schistosomiasis 112
Severe acute respiratory syndrome (SARS) *see* SARS
Singapore: SARS 46
Slash-and-burn farming 114, 126–127
Sleeping sickness 12, 14
Smallpox 40; as biological weapon 23; eradication 13, 21; spread 51; vaccinations 14, 210; vaccine 92; victims 93
Snow, John 20

Social unrest 25, 100–111
South Africa: HIV/AIDS 134
South America: Junin virus 51
Soviet Union: breakup 100–107
Spanish flu 84
Staphylococcus: drug resistance 49
Sudan: Guinea worm disease 9
Syphilis 51

T

Tanzania: refugee camps 108–109
Tetanus 13
Thailand: cervical cancer prevention 172; HIV/AIDS prevention 150–151, 174; insect-borne diseases 71; lifestyle 71; malaria prevention 71
Tick-borne diseases 50–51
Tobacco use 54, 67; public health campaigns 171
Tokyo, Japan 24–25, 50; airport 83
Trachoma 13
Trypanosomiasis *see* Sleeping sickness
Tuberculosis 12, 21, 53, 82, 104; diagnosis 97; drug resistance 49, 104; and HIV/AIDS 134; treatment 106–107

U

Uganda: HIV/AIDS prevention 174; mourning woman 38–39; public health campaigns 174
Undernutrition 10, 12, 30–31
United Kingdom: foot-and-mouth disease 52, 82
United Nations Children's Fund: measles prevention 169–170
United Nations Foundation 168
United States: agriculture 190; border with Mexico 94–95, 96–97; *E. coli* outbreaks 52, 82, 190; foreign aid 8; immigration 94–95, 96–97; influenza 84; life expectancy 230; malaria 112; Navy recruits 68–69; obesity 54; population growth 56, 59; SARS 46; West Nile virus 51
University of California: Lawrence

Berkeley Laboratory 32–33
U.S. Army: medical evacuation team 90–91; Medical Research Institute 40–41

V

Vaccines and vaccinations 7, 210–221; Bangladesh 11, 13, 212; Brazil 132–133; China 169; development 222; DPT 55, 210; hepatitis B 169, 210; HIV/AIDS 210; immunization rates 13; India 168–169, 211–213; influenza 210; measles 13, 14, 169–170, 210, 216–217; polio 11, 55, 168–169, 211–213; tetanus 13; West Nile virus 210, 222; Yellow fever 130, 132–133, 210
Vietnam: SARS 46
Viruses 40, 49, 132; new strains 86

W

War 4, 100, 109, 111
Water: clean 54, 176–183; contamination 53, 187
Waterborne illnesses 20, 57, 100, 171, 176, 182–183; prevention 172
West Nile virus 51; spread 82, 88; transmission 52; vaccines 210, 222
Women: childbirth 10; education 42; empowerment 194–209; loan programs 194, 196–197, 206, 208
World Bank: river blindness prevention 170
World Health Organization: measles prevention 169–170; river blindness prevention 170

Y

Yamomami Indians: malaria 112
Yaws 14
Yellow fever 112, 130; vaccines 130, 132–133, 210

Z

Zaire: diseases 100; unrest 100

IMPACT

Karen Kasmauski and Peter Jaret

Published by the National Geographic Society

John M. Fahey, Jr., President and Chief
    Executive Officer

Gilbert M. Grosvenor, Chairman of the Board

Nina D. Hoffman, Executive Vice President

Prepared by the Book Division

Kevin Mulroy, Vice President and
    Editor-in-Chief

Charles Kogod, Illustrations Director

Marianne R. Koszorus, Design Director

Staff for this Book

Lisa Lytton, Project Editor

K.M. Kostyal, Text Editor

Bill Douthitt, Illustrations Editor

David Griffin, Art Director

David M. Jeffery, Legend Writer

Lewis Bassford, Production Project Manager

Meredith Wilcox, Illustrations Assistant

Judith Klein, Consulting Editor

Katy Hall, Editorial Assistant

Connie Binder, Indexer

Manufacturing and Quality Control

Christopher A. Liedel, Chief Financial Officer

Phillip L. Schlosser, Managing Director

John T. Dunn, Technical Director

Vincent P. Ryan, Manager

Clifton M. Brown, Manager

Library of Congress Cataloging-in-Publication Data available upon request.
ISBN 0-7922-6372-3

One of the world's largest nonprofit scientific and educational organizations, the National Geographic
Society was founded in 1888 "for the increase and diffusion of geographic knowledge." Fulfilling
this mission, the Society educates and inspires millions every day through its magazines, books,
television programs, videos, maps and atlases, research grants, the National Geographic Bee, teacher
workshops, and innovative classroom materials. The Society is supported through membership
dues, charitable gifts, and income from the sale of its educational products. This support is vital to
National Geographic's mission to increase global understanding and promote conservation of our
planet through exploration, research, and education.

For more information, please call 1-800-NGS LINE (647-5463) or write to the following address:
National Geographic Society, 1145 17th Street N.W., Washington, D.C. 20036-4688 U.S.A.
Visit the Society's Web site at www.nationalgeographic.com.